A Comprehensible Guide to J1939

Copperhill Media Corporation
http://www.copperhillmedia.com

A Comprehensible Guide to J1939

By Wilfried Voss

Published by
Copperhill Media Corporation
158 Log Plain Road
Greenfield, MA 01301

ISBN: 978-0-9765116-3-2

Printed in the United States of America

Limit of Liability/Disclaimer of Warranty

158 Log Plain Road
Greenfield, MA 01301

About this book

After writing "A Comprehensible Guide to Controller Area Network", documenting the SAE J1939 standard seemed to be a logical choice when it came to investigating CAN based higher layer protocols. As I have learned from a number or professionals in the CAN industry, J1939 is still gaining enormous popularity, even though it is already in business for some years. However, the quality and availability of documentation on J1939 is in utter contrast to its popularity.

According to Wikipedia, the biggest multilingual free-content encyclopedia on the Internet, "SAE J1939 is the vehicle bus standard used for communication and diagnostics among vehicle components, originally by the car and heavy duty truck industry in the United States." Beyond that statement there is only few information to find on J1939 and the same is true for the entire World Wide Web.

The SAE (Society of Automotive Engineers), like many standardization organizations, is keeping a close lock on their written works and profound information on their web site is sparse. Downloading the J1939 PDF documents comes with a hefty price tag, even for SAE members who receive a discount not worth mentioning.

At the time when this book was released the only available and complete technical reference on J1939 was provided by the SAE either as PDF downloads for a price tag of US$595.00 (Single-User, one-year subscription) or one could buy the complete SAE J1939 standard in one colossal work of 1600+ pages for a mere US$310.00. Regular books dedicated to J1939 and available for a reasonable price did not exist at the time when I started the research and, as mentioned previously, valid references on the World Wide Web were extremely sparse.

Beyond the commercial aspects there is also the issue of educational value and readability of these standards. Standards, like those developed for J1939, are not designed to educate or, God forbid, entertain.

Standardization organizations seem to be the worst when it comes to providing comprehensible and readable documentation of the products they are trying to promote, which ironically creates revenues for technical writers who actually know their job.

Despite the poor condition of the written standard, it was initially a pleasure to investigate the J1939 protocol functions. SAE J1939 is a very ingeniously designed protocol that takes a resourceful advantage of the CAN 29-Bit message identifier. Rather than relying on a myriad of protocol functions, SAE J1939 uses predefined parameter tables, which keeps the actual protocol on a comprehensible level. SAE J1939 is a prime example of good American engineering according to the KISS principle (KISS = Keep It Simple, Stupid!), but it is nevertheless at least as effective as, for instance, CANopen or DeviceNet.

I had originally contemplated to continue my "Comprehensible Guide" series with CANopen, but was overwhelmed by the amount of information that I would need to compile, which also indicates the effort it takes for the newcomer to get familiarized with the topic. SAE J1939 was so much more fun to investigate (again, initially), because it seemed simple and straight-forward. However, this conclusion can only be made after relentless digging through the standards and, after repeated reading, finally understanding what the authors were trying to convey to the reader.

This book is an attempt to create an enjoyable and readable J1939 reference for everybody. The information provided in this book is, besides the SAE J1939 Standards Collection, based on publicly available information such as, but not limited to web sites and printed literature as well as contributions by engineers familiar with Controller Area Network and the J1939 protocol. The information in this book, while based on the J1939 standard, is not a reproduction of any copyrighted SAE document.

Also, this book does not intend to replace the entire SAE J1939 Standards Collection, especially since the standards SAE J1939 and SAE J1939/71 contain mainly data references which account to more than 1000 pages of 8.5 x 11" in size. These data references are not part of this book. The mere intention was to explain the standard in the sense of being a comprehensible guide.

I also need to apologize in advance that the information in this book may seem to be repetitive at times.

First of all, I always try to provide a generic overview of the topic covered in my books. This will help people with a lesser technical background to understand the technology without having to read all details.

Initially I have been trying to describe the collective SAE J1939 Standard by going through the sequence of documents as numbered by the SAE (J1939, J1939/01...J1939/81). As it turns out, a great amount of information in these documents is redundant. It really seems to be the case that individual groups with different interests created their individual documents, sometimes referring to information in other documents, and in other cases reproducing the same information in different form.

In all consequence, the only documents needed to understand the protocol features are:

J1939-21   Data Link Layer
J1939-81   Network Management

The information in this book is based on these documents.

Let me point to the legal disclaimer that states that the publisher and author have used their best efforts in preparing this book.

I would also like to take the opportunity and apologize to all engineers of the SAE who worked on the J1939 standards collection. My comments throughout this book, regarding the condition of the documentation, are not favorable. You have created a great protocol, but the standard is poorly written and lacks any visible structure. Working through the standard was at times tiresome and frustrating.

It was especially irritating to learn that the SAE engineers who created the standard were not fully familiar with the CAN specification. The SAE J1939 Standards Collection contains a number of references to the CAN standard that are misleading in the best case, while others are plain wrong.

One would also expect that engineers, regardless of their special expertise, are familiar with the unit of time, "ms" or "msec" (milli-seconds). Instead the J1939 standard uses mS, which is officially milli-Siemens (electric conductance, equal to inverse Ohm - Ω).

Last, but not least, in case you have questions related to J1939 and/or you would like to contact me, please do so (by any means) through one of my web sites, http://www.J1939Forum.com. Just post an inquiry and either I or the community or both will respond.

About the author

Wilfried Voss is the President of esd electronics, Inc., a company specializing in CAN technology. The company is located in Greenfield, Massachusetts. Mr. Voss has worked in the CAN industry since 1997 and before that was a specialist in the paper industry. He has a master's degree in electrical engineering from the University of Wuppertal in Germany.

Mr. Voss has conducted numerous seminars on CAN and CANopen during various *Real Time Embedded And Computing Conferences* (RTECC), ISA (Instrumentation, Systems, and Automation Society) conferences and various other events all over the United States and Canada. He is also the founder of Copperhill Technologies, a software engineering and consulting company, and the creator of VisualSizer, a comprehensive servo motor sizing software.

Mr. Voss has traveled the world extensively, settling in New England in 1989. He presently lives in an old farmhouse in Greenfield, Massachusetts with his Irish-American wife, their son Patrick and their Rhodesian Ridgeback.

Acknowledgements by the author

This book would not have been possible without the help of my wife, Dr. Susan Marie Voss, a vigorous proof reader and source of many inspirations. Special appreciation is in order for my son Patrick (he was one year old at the release of this book) who taught me how to sufficiently type with only one hand, left or right, while the other was busy keeping him away from the keyboard's *Sleep* button.

A great deal of gratitude shall be attributed to the Boston Red Sox, World Series Champions of 2004 and 2007. Go Sox!

Table of Contents

Introduction to J1939

The Society of Automotive Engineers (SAE) Truck and Bus Control and Communications Subcommittee has developed a family of standards concerning the design and use of devices that transmit electronic signals and control information among vehicle components. SAE J1939 and its companion documents have quickly become the accepted industry standard and the Controller Area Network (CAN) of choice for off-highway machines in applications such as construction, material handling, and forestry machines.

J1939 is a higher-layer protocol based on Controller Area Network (CAN). It provides serial data communications between microprocessor systems (also called Electronic Control Units - ECU) in any kind of heavy duty vehicles. The messages exchanged between these units can be data such as vehicle road speed, torque control message from the transmission to the engine, oil temperature, and many more.

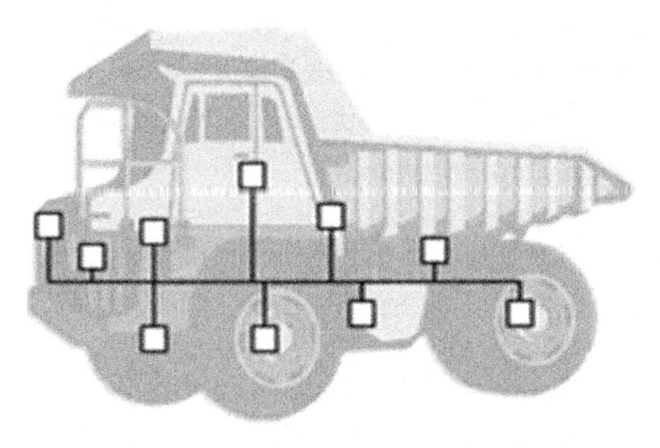

The main advantages of using CAN as a field-bus technology are reduced wiring (CAN requires only two wires between nodes), extremely reliable communication, easy implementation and improved maintenance and service capabilities, which consequently not only produce better vehicle performance, but also help to reduce production costs.

J1939-based protocols are used in:

- Diesel power-train applications
- In-Vehicle networks for trucks and buses
- Agriculture and forestry machinery (ISO 11783)
- Truck-Trailer connections
- Military vehicles (MiLCAN)
- Fleet management systems
- Recreational vehicles
- Marine navigation systems (NMEA2000)

The protocol features of J1939 are based on two older SAE (Society of Automotive Engineers) specifications:

1. SAE J1708

SAE J1708 specifies on the physical layer of the communication link. It uses RS485 as an electrical layer operating at 9600 baud. (Note: Unlike RS232/485 there are no message collisions under CAN). Messages under J1708 start with a Message Identification Character, followed by the data information and a checksum. The message length is 21 characters (or less) and each data character is 10 bits long. Each character starts with a start bit of low polarity.

2. SAE J1587

SAE J1587 is a joint SAE/TMS "Recommended Practices for Electronic Data Exchange Between Microcomputer Systems in Heavy-Duty Vehicle Applications". It regulates the communication and standardized data exchange between ECUs based on J1708 networks.

Note: The situation regarding documents/literature on J1708 and J1587 is as dire as with J1939.

J1939 is designed to replicate the functionality of J1708 and J1587 including control system support. Vehicle applications may utilize either one of both specifications.

The J1939 specification is described by a number of SAE documents, the SAE J1939 Standards Collection:

Document	Description
J1939	Recommended Practice for a Serial Control and Communications Vehicle Network[1]
J1939/01	Recommended Practice for Control And Communications Network for On-Highway Equipment
J1939/02	Agricultural and Forestry Off-Road Machinery Control and Communication Network[2]
J1939/11	Physical Layer - 250k bits/s, Twisted Shielded Pair
J1939/13	Off-Board Diagnostics Connector
J1939/15	Reduced Physical Layer, 250k bits/sec, Un-Shielded Twisted Pair (UTP)
J1939/21	Data Link Layer
J1939/31	Network Layer
J1939/71	Vehicle Application Layer
J1939/73	Application Layer - Diagnostics
J1939/74	Application - Configurable Messaging
J1939/75	Application Layer - Generator Sets and Industrial
J1939/81	Network Management

SAE J1939 is a very ingeniously designed protocol that takes a resourceful advantage of the CAN 29-Bit message identifier. Rather than relying on a myriad of protocol functions, SAE J1939 uses predefined parameter tables, which keeps the actual protocol on a comprehensible level. SAE J1939 is a prime example of good American engineering according to the KISS principle (KISS = Keep It Simple, Stupid!), but it is nevertheless at least as effective as any other higher layer protocol based on CAN.

---

[1] This document does exist, however, the hyperlink on the SAE web site produces an error message.
[2] This document is not listed in the "Core J1939 Standards" list on the SAE web site, but it can be found through their search feature. Reference: http://www.sae.org/technical/standards/J1939/2_200608

Unlike other CAN based protocols J1939 does not implement the established Master/Slave or Client/Server architecture, yet another contribution to the efforts of keeping it simple. The conventional Multi-Master principle (the node that wins the bus arbitration is the master, while all other nodes are the slaves) works just as well.

Overview – Controller Area Network and J1939

The standard CAN message frame uses an 11-bit message identifier (CAN 2.0A), which is sufficient for the use in regular automobiles and any industrial application, however, not necessarily for off-road vehicles.

The Society of Automotive Engineers (SAE) Truck and Bus Control and Communications Subcommittee had developed a family of standards concerning the design and use of devices that transmit electronic signals and control information among vehicle components. As a result, the higher layer protocol SAE J1939, based on CAN, was born, which was required to provide some backward-compatible functionality to older RS485-based communication protocols (J1708/J1587).

In order to serve these demands, the CAN standard needed to be enhanced to support a 29 bit message identifier. The ISO 11898 amendment for an extended frame format (CAN 2.0B) was introduced in 1995.

The 29 bit message identifier consists of the regular 11 bit base identifier and an 18 bit identifier extension. The distinction between CAN base frame format and CAN extended frame format is accomplished by using the IDE bit inside the Control Field. A low (dominant) IDE bit indicates an 11 bit message identifier, a high (recessive) IDE bit indicates a 29 bit identifier.

An 11 bit identifier (standard format) allows a total of 2^{11} (= 2048) different messages. A 29 bit identifier (extended format) allows a total of 2^{29} (= 536+ million) messages.

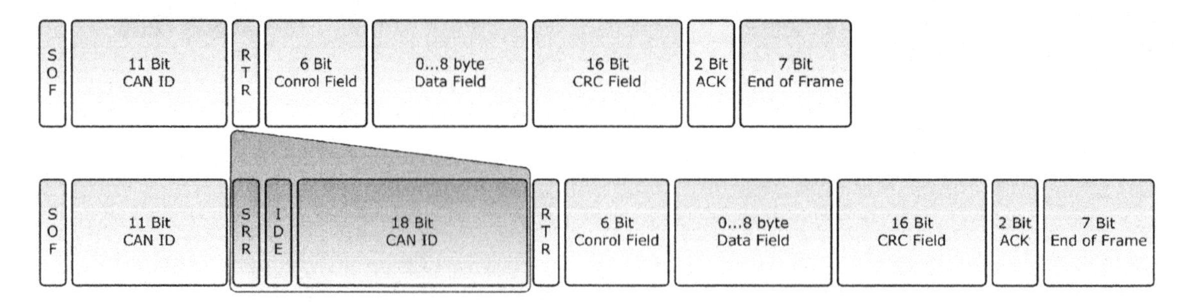

Picture 2.1 Extension from 11-Bit to 29-Bit CAN Identifier

The above picture shows a comparison between a standard CAN data frame with an 11-Bit identifier and a CAN data frame in extended format (29-Bit identifier). Both frames contain an Identifier Extension Bit (IDE)[3], which is at low level for the standard frame and at high for the extended data frame. CAN controllers must be designed in a way that they check the IDE in order to distinguish between the two possible frame formats.

Both formats, Standard (11 bit message ID) and Extended (29 bit message ID), may co-exist on the same CAN bus. During bus arbitration the standard 11 bit message ID frame will always have higher priority than the extended 29 bit message ID frame with identical 11 bit base identifier and thus gain bus access.

The Extended Format has some trade-offs: The bus latency time is longer (minimum 20 bit-times), messages in extended format require more bandwidth (about 20 %), and the error detection performance is reduced (because the chosen polynomial for the 15-bit checksum is optimized for frame length up to 112 bits).

[3] The IDE in an 11-Bit standard frame is embedded in the Control Field.

2.1 CAN Characteristics

Everything that has to do with CAN is based on maximum reliability with the maximum possible performance in mind. After all, CAN was originally designed for automobiles, definitely a very demanding environment for microprocessors, not only in regards to required electrical robustness, but also due to high speed requirements for a serial communication system.

Many companies in the field of medical engineering chose CAN because they have to meet particularly strict safety requirements. Similar problems have been faced by manufacturers of other equipment with very high safety or reliability requirements, including robots, lifts and transportation systems.

The CAN properties can be summarized as:

- Multi-Master priority based bus access
- Non-destructive contention-based arbitration
- Multicast message transfer by message acceptance filtering
- Remote data request
- Configuration flexibility
- System-wide data consistency
- Error detection and error signaling
- Automatic retransmission of messages that lost arbitration
- Automatic retransmission of messages that were destroyed by errors
- Distinction between temporary errors and permanent failures of nodes
- Autonomous deactivation of defective nodes

2.1.1 Frames

In the language of the CAN standard, all messages are referred to as frames, such as data frames, remote frames, error frames, etc. Information sent to the CAN bus must be compliant to defined format frames of different but limited length.

Any node connected to the network may transmit a new frame as soon as the bus is idle.

The consistency of a frame must be simultaneously accepted by all nodes in a CAN network.

CAN provides four different types of message frames:

- **Data Frame – Sends data**
  Data transfer from one sending node to one or numerous receiving nodes.
- **Remote Frame - Requests data**
  Any node may request data from another source node. A remote frame is consequently followed by a data frame containing the requested data.
- **Error Frame - Reports error condition**
  Any bus participant, sender or receiver, may signal an error condition at any time during a data or remote frame transmission.
- **Overload Frame - Reports node overload**
  A node can request a delay between two data or remote frames, meaning the overload frame can only occur between data or remote frame transmissions. Considering today's technologies, Overload frames should not occur in a properly functioning network. As a matter of fact, a number of modern CAN controllers does not even support an Overload frame anymore.

The Remote Transmission Request is not available for use in SAE J1939, since it uses Acknowledgement messages. In addition, in August of 2005 CAN-in-Automation (CiA) had released (but not promoted) an application note 802 - "CAN remote frame - Avoiding of usage", where the use of Remote Frames is disputed.

2.1.1.1 CAN Data Frame Architecture

Per CAN standard a data frame consists of the following components:

- **SOF** (Start of Frame) - Marks the beginning of data and remote Frames
- **Arbitration Field** – Includes the message ID and RTR (Remote Transmission Request) bit, which distinguishes data and remote frames
- **Control Field** – Used to determine data size and message ID length
- **Data Field** – The actual data (Applies only to a data frame, not a remote frame)
- **CRC Field** - Checksum
- **EOF** (End of Frame) – Marks the end of data and remote frames

Picture 2.1.1.1.1 shows a detailed view of the CAN data frame architecture:

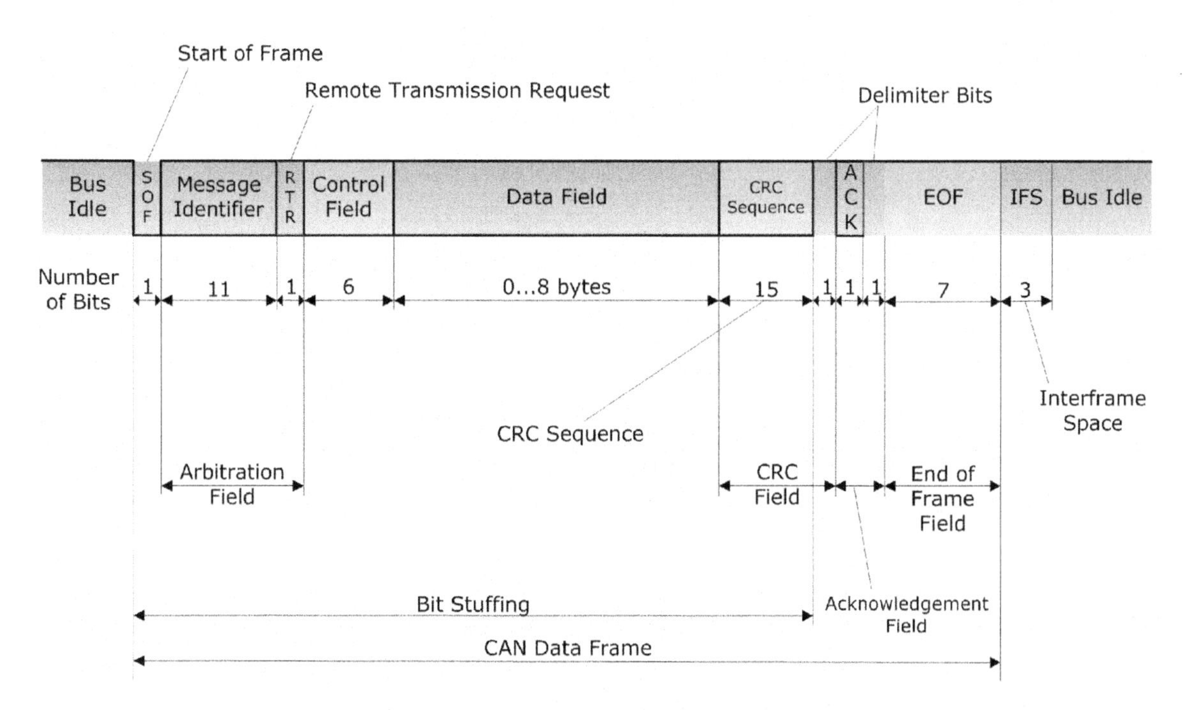

Picture 2.1.1.1 CAN Data Frame Architecture

2.1.2 Multi-Master Bus Access

In order to assure a direct communication between nodes, therefore providing maximum speed combined with maximum reliability, CAN does not restrict itself to the popular client/master network configuration. In a typical CAN network all nodes have equal rights. Every node transmitting a data/remote frame will be the bus master during that transmission.

The actual bus access is managed through non-destructive bit-wise arbitration, which in turn provides very effective message collision avoidance in case that multiple nodes attempt to access the bus at the same time.

A possible bus access conflict is being resolved through contention-based arbitration using the message identifier. The CAN arbitration process assures in a very timely fashion that no information will be lost. The transmitter with the frame of highest priority (lowest message ID) will gain the bus access.

Frames that lost the arbitration and frames that were interrupted by error conditions will be retransmitted automatically as soon as the bus is idle again.

2.1.3 Message Broadcasting

The broadcasting of messages is based on a producer-consumer principle. One node, when sending a message, will be the producer while all other nodes are the consumers. All nodes in a CAN network receive the same message at the same time.

In a multi-master network nodes may transmit data at any time. Each node "listens" to the network bus and will receive every transmitted message. The CAN protocol supports message-filtering , i.e. the receiving nodes will only react to data that is relevant to them.

Messages in CAN are not confirmed because that would unnecessarily increase the bus traffic. CAN assumes that all messages are compliant with the defined standard and if they are not, there will be a corresponding response by all nodes in the network. All receiving nodes check the consistency of the received frame and acknowledge the consistency. If the consistency is not acknowledged by any or all nodes in the network, the transmitter of the frame will post an error message to the bus.

If either one or more nodes are unable to decode a message, i.e. either detect an error in the message or are unable to read the message due to an internal malfunction, the entire bus will be notified of the error condition. Nodes that transmit faulty data or nodes that are constantly unable to receive a message correctly remove themselves from the bus and thus allow restoration of proper bus conditions.

The CAN standard also defines the request of a transmission from another node by means of a remote frame. The remote frame and the requested data frame use the same message identifier. They are, however, distinguished by the RTR bit (Remote Transmission Request) during the arbitration process.

2.1.4 Message Priority

Per definition, CAN nodes are not concerned with information about the system configuration (e.g. node address, etc.), hence CAN does not support node IDs. Instead, receivers process messages by means of an acceptance filtering process, which decides whether the received message is relevant for node's application layer or not. There is no need for the receiver to know the transmitter of the information and vice versa.

CAN data transmissions are distinguished by a unique message identifier (11/29 bit), which also represents the message priority. A low message ID represents a high priority. A single CAN node may send or receive any number of messages, which contributes, yet again, to a maximum level of flexibility.

High priority messages will gain bus access within shortest time even when the bus load is high due to the number of lower priority messages.

Message transmissions are usually event-driven to reduce the bus load and that guarantees short latency times for real-time applications.

2.1.5 Short Messages

CAN supports messages between 0 and 8 bytes of length. Initially this may seem to be a disadvantage compared to other technologies, but a short data length also assures short latency times for high priority messages. Additionally a limited data length contributes to the ability to withstand the strain of harsh electrical environments.

A message length of a maximum of 8 bytes is sufficient for data communications in cars, smaller machines such as household appliances and lower level automation. Higher-Layer protocols such as CANopen support segmented transmission of data of unlimited length that are more suitable for complex automation tasks such as motion control. However, CANopen still supports the limited 8 byte messages (Process Data Objects = PDO). These messages are usually assigned a high priority and, according to the CAN standard, they are able to interrupt the segmented transfer of low priority data (Service Data Object = SDO) after each completed data segment of 8 bytes. The low priority data transfer will resume right after the high priority message has been transmitted.

2.1.6 Bus Arbitration

Since a serial communication system such as CAN is based on a two-wire connection between nodes in the network, i.e. all nodes are sharing the same physical communication bus, a method of message/data collision avoidance is mandatory to assure a safe data transfer and to avoid delays resulting from the necessary restoration of proper bus conditions after the collision.

A collision may occur when two or more nodes in the network are attempting to access the bus at virtually the same time, which may result in unwelcome effects, such as bus access delays or even destruction or damage of messages.

CAN provides a non-destructive bus arbitration, i.e. no message gets lost. Higher priority messages will win the bus access, while low priority message wait until their time has come. Based on a 1 MBit/sec baud rate and an 11 Bit message identifier, the arbitration process is finished after 12 microseconds.

2.1.7 Error Detection and Fault Confinement

Rather than providing a message confirmation, which in turn would increase the bus load, CAN goes the more aggressive route of assuming that all messages must be consistent with the defined standard. Every diversion from this standard is detected and reported immediately, i.e. the error detection actually replaces the message confirmation. Naturally, confirmed messages would occur more often than actual error messages.

Each node in the network will receive each transmitted message. A message filter guarantees that the node knows when to ignore a message or to process it. However, each node in the network will check the transmitted message for compliance with the defined standard. All receiving nodes check the consistency of the received frame and acknowledge the consistency. If the consistency is not acknowledged by any or all nodes in the network, the transmitter of the frame will post an error frame to the bus.

The occurrence of an error frame may actually have two reasons. First, the transmitted data frame was really faulty or second, the data frame was correct, but one node erroneously reported an error due to a local reception problem.

It is important to distinguish between temporary errors or permanent failures of a node. CAN controllers address this problem by providing two different error counters, one for transmit errors and one for receive errors. If either counter exceeds a certain limit, the node is considered faulty.
As part of the fault confinement, the CAN protocol allows the "removal" of a CAN node from the network, in case the node produces a constant stream of errors and therefore unnecessarily increases the bus load.

CAN also provides very short error recovery times of a maximum of 23 bit times. With a baud rate of 1 MBit/sec this translates into maximal 23 microseconds.

For further, more detailed information on Controller Area Network (CAN) please refer to "A Comprehensible Guide to Controller Area Network" from the same author and publisher (http://www.copperhillmedia.com).

2.2 CAN Higher Layer Protocols

Even though extremely effective in automobiles and small applications, CAN alone is not suitable for machine automation, since its communication between devices is limited to only 8 bytes per message. As a consequence, higher layer protocols such as CANopen for machine control, DeviceNet for factory automation and J1939 for vehicles were designed to provide a real networking technology that support messages of unlimited length and allow a master/slave configuration.

In order to explain higher layer protocol we must refer to the ISO/OSI 7-Layer Reference Model[4] as shown in the picture below.

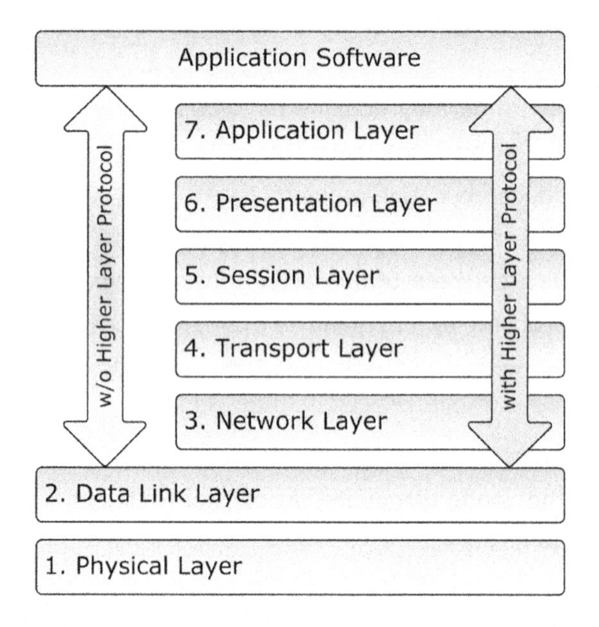

Picture 2.2.1 ISO/OSI 7-Layer Reference Model

The standard CAN implementation bypasses the connection between the Data Link Layer and the Application Layer. The layers above the Data Link Layer are covered by additional software, which represents per definition a higher layer protocol.

[4] For more information on the OSI Reference Model refer to:
CertificationZone.com OSI Reference Model Pocket Guide
by Howard C. Berkowitz - ISBN: 1890911143

Be aware, whenever you attempt to add software functions between the CAN Data Link Layer and the Application Layer, you will be adding functionalities that are already covered by off-the-shelf available higher layer protocols such as CANopen, DeviceNet and J1939.

To put it in a nut shell, higher layer protocols are necessary, because

- They enable data transport of more than 8 bytes per message
- Embedded Systems may require an appropriate communication model based on Master/Slave configuration
- They provide Network Management (Network Start-Up, Node Monitoring, Node Synchronization, etc.)

The most popular higher layer protocols based on Controller Area Network are:

- CANopen
- DeviceNet
- SAE J1939[5]

CANopen and DeviceNet are mainly used in industrial control applications. CANopen, however, is to a certain degree also suitable for vehicle applications. Both protocols provide powerful and complete protocol features, but it is also safe to say that they are over-engineered, which makes it very difficult to understand them in their entirety.

SAE J1939 is a very ingeniously designed protocol that takes a resourceful advantage of the CAN 29-Bit message identifier. Rather than relying on a myriad of protocol functions, SAE J1939 uses predefined parameter tables, which keeps the actual protocol on a comprehensible level. SAE J1939 is a prime example of good American engineering according to the KISS principle (KISS = Keep It Simple, Stupid!), but it is nevertheless at least as effective as CANopen or DeviceNet.

[5] Per standard J1939 does not support a Master/Slave network configuration or Network Management features at the same level of CANopen and DeviceNet, which does, nevertheless, not mean that such functions cannot be implemented at the application level.

2.2.1 CANopen

- Is suited for embedded applications
- Was originally designed for motion control
- Was developed and is maintained by the CAN-in-Automation User Group

Like CAN, the CANopen standard is the responsibility of CiA (CAN-in-Automation). For further information, refer to http://www.can-cia.org.

2.2.2 DeviceNet

- Is suited for industrial applications (floor automation)
- Was developed by Allen Bradley/Rockwell
- Is maintained by Open DeviceNet Association (ODVA)

The DeviceNet Specification, consisting of two volumes: Volume One - Common Industrial Protocol (CIP) and Volume Three- DeviceNet Adaptation of CIP, is available only for ODVA (Open DeviceNet Vendor Association) members. For further information, refer to http://www.odva.org.

2.2.3 SAE J1939

- Defines communication for vehicle networks (trucks, buses, etc.)
- Does not support a Master/Slave Configuration
- Is a standard developed by the Society of Automotive Engineers (SAE)

The SAE J1939 Standards Collection can be found exclusively on the Web at http://www.sae.org.

2.3 J1939 Characteristics

J1939 is a higher-layer protocol based on Controller Area Network (CAN). It provides serial data communications between microprocessor systems (also called Electronic Control Units - ECU) in any kind of heavy duty vehicles.

Everything that has to do with CAN is based on maximum reliability with the maximum possible performance in mind, not only in regards to required electrical robustness, but also due to high speed requirements for a serial communication system.

While CAN itself is sufficiently suited for communication in a regular automobile or in small industrial applications, it comes with a few short-comings in regards to network management. In order to add these features CAN as the physical layer (the entire CAN protocol is on silicon) can be extended by additional software, the so-called higher layer protocols (such as J1939).

J1939 takes advantage of CAN features such as:

- Maximum reliability
- Excellent error detection & fault confinement
- Collision-free bus arbitration

Other than CAN, which supports up to 1 Mbit/sec, J1939 limits itself to 250 kbit/sec. CAN was designed to be as close to real-time applications as possible. This level of performance is not required for J1939.

It is troubling to learn that the SAE J1939 standard has no problem with compromising the CAN standard. The Network Management (SAE J1939/81), for instance, allows scenarios where two CAN nodes with the same message ID can access the bus. The result of such a situation is unpredictable. In addition, the SAE J1939 message format (as described in SAE J1939/21) does not take advantage of the message filtering as provided by all CAN controllers in the industry.

2.3.1 J1939 Quick Reference

- Higher-Layer Protocol using CAN as the physical layer
- Shielded twisted pair wire
- Max. network length of 40 meters (~120 ft.)
- Standard baud rate of 250 kBit/sec
- Max. 30 nodes (ECUs) in a network
- Max. 253 controller applications (CA) where one ECU can manage several CAs
- Peer-to-peer and broadcast communication
- Support for message length up to 1785 bytes
- Definition of Parameter Groups (Predefined vehicle parameters)
- Network Management[6] (includes address claiming procedure.

It must be emphasized that the maximum network length of 40 m (roughly 120 ft.), the baud rate of 250 kBit/sec and the maximum number of nodes (30) are self-inflicted restrictions by the SAE, most probably with the intention to keep everything on the extreme safe side and thus trying to prevent potential runtime problems.

In all consequence, the network length at 250 kBit/sec, according to ISO 11898, is 250 m (roughly 750 ft.).

There is no reason to believe that J1939 cannot be operated at the max. CAN baud rate of 1 MBit/sec. Naturally, the network length would drop, but the mere J1939 protocol features post no restriction in regards to the baud rate.

The J1939 protocol utilizes an 8 bit device (ECU) address, which would allow the operation of 256 nodes in the same network. It can only be assumed that the SAE was trying to keep the bus traffic on a low level by restricting the maximum number of nodes to 30. Elaborating comments on this restrictions may be embedded somewhere in the standard.

[6] The SAE J1939 Network Management does not include support for a Master/Slave configuration and it does not include node monitoring. These functions can nevertheless be implemented on an application level.

In all consequence the ECU address is really a Controller Application address in a situation where each ECU may accommodate several Controller Applications. The 253 addresses (Address 254 is reserved for Network Management, Address 255 is used for global addressing, i.e. message broadcasting) are assigned (claimed) for the Controller Applications, not the actual ECU.

The following picture shows an example, where, for instance, ECU A accommodates three controller applications.

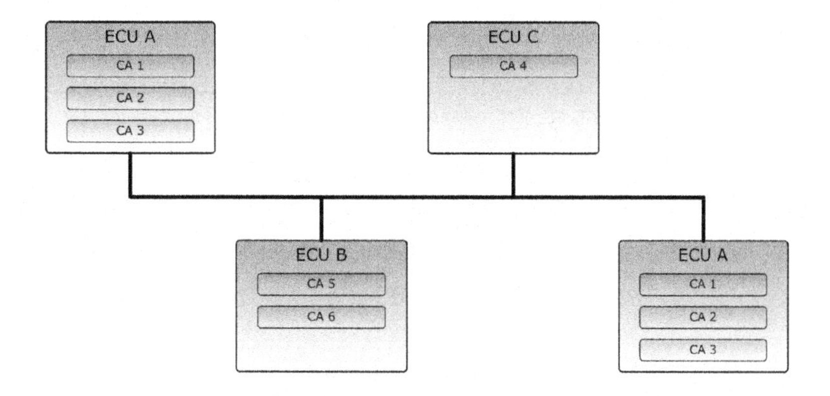

Picture 2.3.1.1 Sample J1939 Network

The picture also demonstrates that ECUs of the same function (ECU A) can co-exist in a J1939 network without address collision. J1939 features a very ingenious feature, the Address Claim procedure which automatically assigns addresses to each Controller Application. In case of an Address Claim conflict, the Controller Applications are able to claim another free address.

The main characteristic of J1939 is, however, the use of Suspect Parameter Numbers (SPN) and Parameter Group Numbers (PGN) which point to a huge set of predefined vehicle data and control functions. The definition of PGNs and Suspect Parameter Numbers (SPN) make out the bulk of the SAE J1939 Standards Collection. SPNs define the data type and use for each parameter in a parameter group. The Parameter Group Number is embedded in the 29-Bit CAN message identifier.

The following picture demonstrates the use of Suspect Parameter Numbers, Parameter Groups and Parameter Group Numbers.

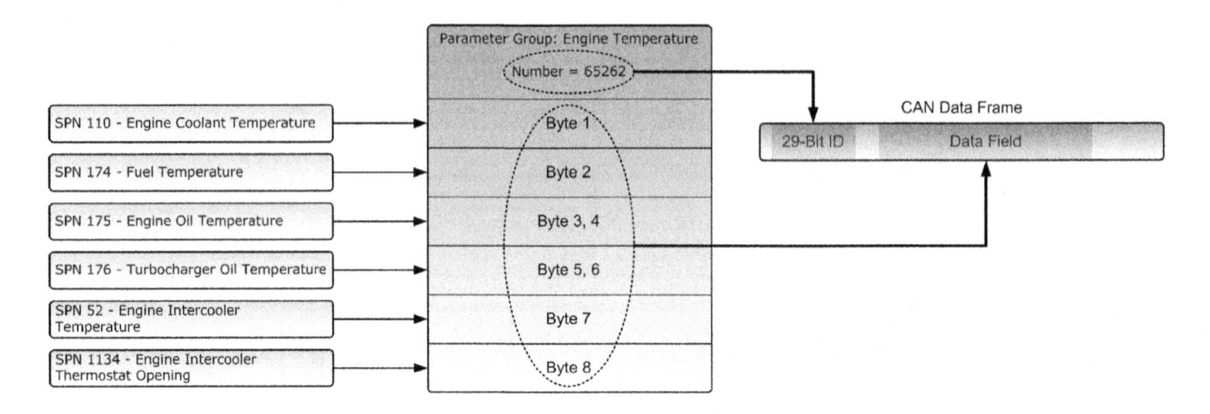

Picture 2.3.1.2 SPNs and PGNs

The Suspect Parameter Numbers in this example point to vehicle data associated with the Engine Temperature. Note that all SPNs have been selected logically to fit into the Parameter Group *Engine Temperature*.

The Parameter Group *Engine Temperature* has been assigned the Parameter Group Number (PGN) 65262. It is in fact the Parameter Group Number (embedded in the 29-Bit message ID) plus the actual data (embedded in the CAN data field) that is being transmitted into the CAN bus.

It is a requirement that all nodes sending or receiving the PGN 65262 know the structure of the PGN as well as all associated SPNs.

2.3.2 J1939 Message Format

CAN supports 11- and 29-Bit message identifiers. CAN is also designed in a way where the sending node is not concerned with which node(s) receive(s) the data. In turn a receiving node does not know/care who sent the data.

In contrast J1939 uses only the 29-Bit identifier (In fact, the CAN standard was extended from 11 to 29 bit per request by the SAE in order to support J1939). J1939 uses the identifier, among other features, to identify the source and, in some cases, the destination of data on the bus.

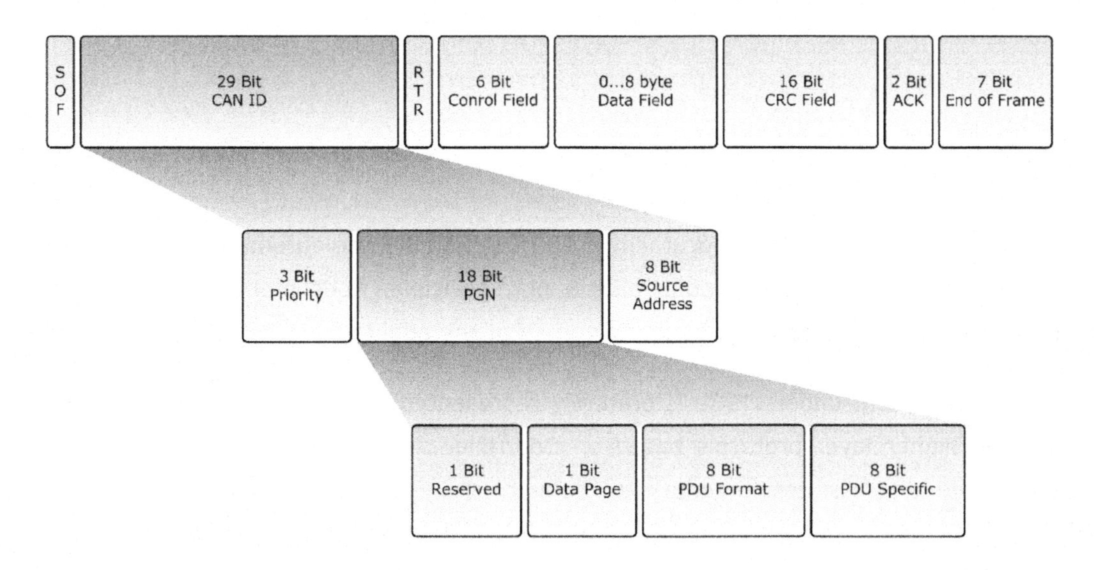

Picture 2.3.2.1 J1939 Message Format

As shown in the picture J1939 extends the use of the 29-Bit CAN identifier beyond the standard CAN message identification.

The CAN identifier is split into a priority field, a Parameter Group Number (PGN)[7] to identify the content of the data field, and the source address.

A message priority '0' indicates highest priority and a message priority of '7' indicates lowest priority. High priorities are usually assigned to time-critical data such as torque control

[7] For a peer-to-peer communication the PDU Specific field is used as destination address and, in this case, is not part of the PGN. See also chapter *Parameter Group Number Architecture*.

messages to the engine, while lower priorities indicate not time critical messages such as vehicle road speed.

Messages are being transmitted according to the producer/consumer model (broadcast) or destination specific (peer-to-peer).

The most important feature under J1939, though, is represented by the Parameter Group Numbers (PGN) that point to Parameter Groups (PG). Parameters groups are, for instance, engine temperature which includes coolant temperature, fuel temperature, oil temperature, etc. Parameter Groups and their numbers (PGN) are listed in SAE J1939 (roughly 300 pages) and defined in SAE J1939/71, a document containing roughly 800 pages filled with parameter group definitions plus suspect parameter numbers (SPN)[8]. SPNs define the data structure of each entry inside a Parameter Group.

2.3.3 Network Management

The network management (SAE document J1939/81) handles the automatic allocation of node addresses very similar to the popular plug & play principle. Node monitoring is not defined under J1939 and, if needed, must be implemented at the application level.

Network Management under J1939 is primarily represented by the Address Claiming Process. While other higher layer protocols based on Controller Area Network (CAN) do not support dynamic node address assignments per default, the SAE J1939 standard provides this ingeniously designed feature to uniquely identify ECUs and their primary function.

SAE J1939 does not support a conventional Master/Slave or Client/Server network management per default. The flexible structure of the protocol, however, does allow that any additional network management functions can be implemented at the application level.

[8] The assignment of Parameter Groups and associated Parameter Group Numbers and Suspect Parameter Numbers is beyond the scope of this book.

2.4 Other J1939 Based Protocols

Per definition, SAE J1939 provides serial data communications between microprocessor systems (also called Electronic Control Units - ECU) in any kind of heavy duty vehicles. The messages exchanged between these units can be data such as vehicle road speed, torque control message from the transmission to the engine, oil temperature, and many more.

SAE J1939 and its companion documents have quickly become the accepted industry standard of choice for off-highway machines. It was all too natural that organizations and manufacturers in the agricultural, military and marine industries, rather than re-inventing the wheel, adopted the proven combination of physical layer, Controller Area Network (CAN), and J1939 as the higher layer protocol for vehicles. However, it is in the specific nature of agricultural and military as well as marine applications that slight modifications, including a name change, were necessary.

These "new" protocols are[9]:

- ISO 11783 (a.k.a ISOBUS) – Agricultural Industry
- MilCAN – Military Applications
- NMEA 2000 – Marine Applications

[9] Note by the author: While information on NMEA 2000 and especially MilCAN is easily available, documentation of the ISO 11783 standard is "protected" by excessive price tags. One cannot deny the standardization organization's objective to produce revenues, but one also needs to seriously question their marketing expertise.

2.4.1 ISO 11783 (ISOBUS)

ISO 11783, a.k.a. ISOBUS, is a CAN (Controller Area Network) Higher Layer Protocol based on the SAE J1939 standard for forestry and agricultural vehicles. ISO 11783 was a joint development by manufacturers in the agricultural and forestry industry to address the increasing needs for electronic control in the machinery and vehicles they produce.

ISO 11783 consists of the following parts, under the general title Tractors and machinery for agriculture and forestry - Serial control and communications data network:

- Part 1: General standard for mobile data communication
- Part 2: Physical layer
- Part 3: Data link layer
- Part 4: Network layer
- Part 5: Network management
- Part 6: Virtual terminal
- Part 7: Implement messages applications layer
- Part 8: Power train messages
- Part 9: Tractor ECU
- Part 10: Task controller and management information system data interchange
- Part 11: Mobile data element dictionary
- Part 12: Diagnostic
- Part 13: File Server

The ISO 11783 standards can be purchased through the International Organization for Standardization (ISO)[10]. The standard is managed by the ISOBUS group in VDMA (http://www.isobus.net)[11].

[10] The price tags for each document are extraordinary.

[11] The web site lacks the substance to be taken seriously. The bulk of the little information that exists is based on marketing material and most documents are in German. Technical information is virtually non-existent.

2.4.2 MilCAN

According to the official web site (http://www.milcan.org): "MilCAN has been defined by a group of interested companies and government bodies associated with the specification, manufacture and test of military vehicles. The MilCAN working group was formed in 1999 as a sub-group of the International High Speed Data Bus - Users Group (IHSDB-UG) when a need was recognised to standardise the implementation of CANbus within the military vehicles community. The mission statement of this newly formed group was 'To develop, for various application classes in all military vehicles, a common interface implementation specification based on CANbus'."

Describing the MilCAN standard is not an easy task and the only reason it found its way into this book is due to the fact that it is partly based on J1939. It seems that the creators of the protocol tried to satisfy the protective demands of every European member (in this case especially the Germans and Brits) on one side and American companies on the other. One can only appreciate that the circle of members was not extended any further. MilCAN is an inconsistent mixture of CUP, a protocol developed by the German Army (Bundeswehr), SAE J1939, representing the American side, and CANopen, representing the European side.

As a resullt, there are two variants of MilCAN, MilCAN A and MilCAN B. MilCAN A is based on the 29-bit CAN identifier according to SAE J1939, the major difference being that MilCAN A supports deterministic data transfer and accommodates both, synchronous and asynchronous, data. MilCAN B, on the other hand, is based on the 11-bit CAN identifier and can, at least officially, make use of devices that have been designed for CANopen[12]. Also officially, it should be possible to mix MilCAN A (J1939) devices with MilCAN B (CANopen) devices on the same bus.

[12] CANopen devices must be segmented via a bridge.

2.4.3 NMEA 2000

Of all the SAE J1939 derivatives, NMEA 2000 seems to be the only consequent and straight-forward adaptation of J1939. While taking advantage of a proven and ingeniously designed protocol, NMEA 2000 defines only its own messages.

NMEA 2000 is used for marine data networks providing communication between marine specific electronic devices such as depth finders, chartplotters, navigation instruments, engines, tank level sensors, and GPS receivers.

It has been defined and is controlled by the US based National Marine Electronics Association (NMEA). Information on their official web site (http://www.nmea.org) is somewhat sparse. Another web site, http://www.jackrabbitmarine.com, however, provides in-depth information.

NMEA 2000 is a modernized version and replacement of an older standard, NMEA 0183. It has a significantly higher data rate (250k bits/second vs. 4.8k bits/second for NMEA 0183). It also uses a binary message format as opposed to the ASCII serial communications protocol used by NMEA 0183. Another distinction between the two protocols is that NMEA 2000 is a multiple-talker, multiple-listener data network whereas NMEA 0183 is a single-talker, multiple-listener serial communications protocol.

The J1939 Standards Collection

The "SAE Truck and Bus Control & Communications Network Standards Manual - 2007 Edition" is a colossal work of roughly 1600 pages where about 1000 pages refer to data such as Parameter Group Assignments, Address and Identify Assignments, Parameter Group Numbers, and more.

While it combines all standards in one work, it seems that the individual standard descriptions were developed independently from each other. There is no noticeable system to standardize the standards. Most documents provide a table of contents, some of them do not. Some standards repeat information from others or refer to others; many references actually lead nowhere. The page numbering is also inconsistent with the table of contents that exist.

It is apparent throughout the entire J1939 standards collection that the writers/creators of the standard indulged themselves with sophisticated, but nevertheless irritating terms and language.

The J1939 Standards Collection was created for mere documentation purposes without the necessary passion for the subject and without reader-friendliness or educational value in mind. Especially, unnecessary abbreviations (and their abbreviations) are used to an excessive degree and make reading these documents a demanding task.

3.1 ISO/OSI 7-Layer Reference Model

The J1939 Standards Collection was designed to follow the ISO/OSI 7-Layer Reference Model as far as necessary[13]. Each layer is addressed by a corresponding document.

7. Application Layer	J1939/7x
6. Presentation Layer	J1939/6x
5. Session Layer	J1939/5x
4. Transport Layer	J1939/4x
3. Network Layer	J1939/3x
2. Data Link Layer	J1939/2x
1. Physical Layer	J1939/1x

Picture 3.1.1 SAE J1939 Standards

The Open Systems Interconnection Basic Reference Model or OSI Model for short is a layered, abstract description for communications and computer network protocol design.

[13] For more information on the OSI Reference Model refer to:
CertificationZone.com OSI Reference Model Pocket Guide
by Howard C. Berkowitz - ISBN: 1890911143

Layer	Title	Description
7	Application	Supports application and end-user processes.
6	Presentation	Provides independence from differences in data representation (e.g., encryption) by translating from application to network format, and vice versa.
5	Session	Establishes, manages and terminates connections between applications.
4	Transport	Provides transparent transfer of data between end systems, or hosts, and is responsible for end-to-end error recovery and flow control. Ensures complete data transfer.
3	Network	Provides switching and routing technologies, creating logical paths, known as virtual circuits, for transmitting data from node to node.
2	Data Link	Data packets are encoded and decoded into bits. It furnishes transmission protocol knowledge and management and handles errors in the physical layer, flow control and frame synchronization. The data link layer is divided into two sub layers, the Media Access Control (MAC) layer and the Logical Link Control (LLC) layer. The MAC sub layer controls how a computer on the network gains access to the data and permission to transmit it. The LLC layer controls frame synchronization, flow control and error checking.
1	Physical	Conveys the bit stream (electrical impulse, light or radio signal) through the network at the electrical and mechanical level.

In a CAN (Controller Area Network) network both layers, Data Link and Physical Layer, are represented by the actual CAN controller. As a matter of fact, the actual CAN protocol, i.e. the entire data communication management including bus arbitration, error detection and fault confinement, etc., etc., is implemented into silicon. CAN controllers know per default what to do and how to do it.

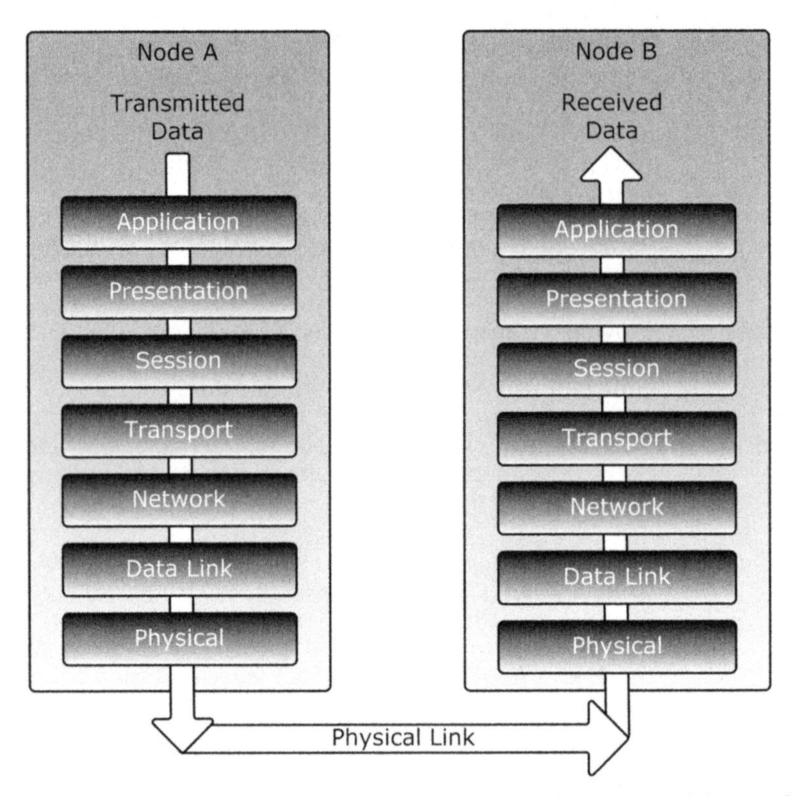

Picture 3.1.2 Node Communication based on 7-Layer OSI Model

Picture 3.1.2 demonstrates the application of the OSI 7-Layer Model to establish the communication between two nodes.

3.2 J1939 Documents

The J1939 Standards Collection was designed to follow the ISO/OSI 7-Layer Reference Model as far as necessary. Each layer is addressed by a corresponding document.

The SAE has named documents addressing the transport (4), session (5), and presentation (6) layer in the ISO/OSI 7-Layer Reference Model. These layers are, however, not documented (in all consequence they are not necessary for J1939) and thus the corresponding documents have not been created.

Additional documents with a /0x suffix, such as J1939/01 and J1939/02, identify the industry or application. J1939/81 does not refer directly to any of the seven layers and is regarded as an additional function affecting all layers.

As the name, J1939 Standards Collection, implies the J1939 standard comprises of a number of documents:

Document	**Description**
J1939	Recommended Practice for a Serial Control and Communications Vehicle Network[14]
J1939/01	Recommended Practice for Control And Communications Network for On-Highway Equipment
J1939/02	Agricultural and Forestry Off-Road Machinery Control and Communication Network[15]
J1939/11	Physical Layer - 250k bits/s, Twisted Shielded Pair
J1939/13	Off-Board Diagnostics Connector
J1939/15	Reduced Physical Layer, 250k bits/sec, Un-Shielded Twisted Pair (UTP)
J1939/21	Data Link Layer
J1939/31	Network Layer
J1939/71	Vehicle Application Layer
J1939/73	Application Layer - Diagnostics
J1939/74	Application - Configurable Messaging
J1939/75	Application Layer - Generator Sets and Industrial
J1939/81	Network Management

[14] This document does exist, however, the hyperlink on the SAE web site produces an error message.

[15] This document is not listed in the "Core J1939 Standards" list on the SAE web site, but it can be found through their search feature. Reference: http://www.sae.org/technical/standards/J1939/2_200608

During researching the Internet (Google Search) references to further documents were found on various web sites. They are, however, not available through the SAE. A search on their web site did not produce any results.

J1939/12	Physical Layer - Twisted Quad of Wires
J1939/14	DIN 9684 - 50k bits/sec[16]
J1939/72	Virtual Terminal (DIN 9684)[17]

In all consequence, the only documents needed to understand the protocol features are:

J1939/21	Data Link Layer
J1939/81	Network Management

The information in this book is based on these documents.

3.2.1 SAE J1939

The full name of the document is "SAE J1939 Recommended Practice for a Serial Control and Communications Vehicle Network"[18]. This document represents the top level of a number of documents known as the J1939 Standards Collection. It references to the OSI 7-Layer Model and the corresponding documents and provides a "J1939 Tutorial". The bulk of the document, though, is comprised of 300+ pages containing data such as Parameter Group Assignments (Appendix A), Address and Identity Assignments (Appendix B) and Fault Reporting Parameters (Appendix C). Without these appendixes the document is a mere 19 pages long, where the first 7 pages are filled the table of content, definitions and abbreviations. In addition, a great amount of information, especially about the J1939 message format, is repeated in different form in SAE J1939/21. Some references to the J1939 Network Management are in stark contrast to SAE J1939/81.

The only reason to justify the existence of this document lies in the fact that it does reference all Parameter Group Numbers (PGN) and Suspect Parameter Numbers (SPN), because their

[16] Source: CAN-in-Automation (CiA). DIN is a German standardization system and the connection to the SAE J1939 standard is not clear at this time.
[17] Source: CAN-in-Automation (CiA). DIN is a German standardization system and the connection to the SAE J1939 standard is not clear at this time.
[18] This document does exist, however, the hyperlink on the SAE web site produces an error message.

definition will be in one of several other documents, which may keep the reader searching for the needle in the haystack.

3.2.2 SAE J1939/11 Physical Layer

The physical layer of J1939 is based on Controller Area Network (CAN) as described in ISO11898. J1939-11 specifies a shielded twisted pair wire with a maximum length of 40 m (roughly 120 ft.). The CAN baud rate is 250 Kbit/sec per standard. The maximum number of ECUs is limited to 30 for one segment. Several units may be connected using special interconnection ECUs.

SAE J1939/11 is an abbreviated version of ISO11898 or the Bosch CAN specification, respectively, with some added electrical specifications for use in vehicles. As a matter of fact, the CAN specifics have been wrecked to a useless degree.

3.2.3 SAE J1939/13 Off-Board Diagnostic Connector

J1939/13 defines a standard connector for diagnostic purpose. The connector is a Deutsch HD10 - 9 – 1939 (9 pins, round connector).

3.2.4 SAE J1939/15 Reduced Physical Layer

J1939/15 describes a physical layer that utilizes an Unshielded Twisted Pair (UTP) cable.

3.2.5 SAE J1939/21 Data Link Layer

J1939/21 defines the use of the CAN data frame (29-bit identifier, Parameter Group Numbers – PGN, etc.) and the transport protocol functions, i.e. a definition of how messages longer than the standard CAN data length (8 bytes) are transmitted in a J1939 bus network[19].

As in the SAE J1939/11 standard any references to CAN specifics have been wrecked to a useless degree.

Without further ado, the J1939/21 standard, assuming that the reader has already studied all applicable publications and knows what Parameter Groups are[20], jumps immediately into the detailed architecture of Parameter Group Numbers. It is like explaining the function of an automobile by starting with the details of the gas injection system. The document is poorly written and lacks any structure.

For instance, the mere fact that some J1939 messages may be longer than the standard CAN message of 8 bytes is mentioned in numerous chapters without offering any substantial details on how the messages are packetized. In the same spirit, topics like, for instance, PDU Formats are explained repeatedly, spanning over several chapters.

3.2.6 SAE J1939/31 Network Layer

J1939/31 describes the services and functions needed for intercommunication between different segments of a J1939 network by means of bridges, routers, gateways, and repeaters.

[19] This statement is based on SAE claims. In all consequence the document offers less than one page on the packaging of messages longer than 8 bytes.
[20] As a matter of fact, the document does not provide any explanation of what Parameter Groups are. The same is true for Suspect Parameter Numbers (SPN)

3.2.7 SAE J1939/71 Vehicle Application Layer

J1939/71 describes and defines the Parameter Group Numbers and Suspect Parameter Numbers. This document is updated very frequently to incorporate new standard parameters and messages.

Parameters groups are, for instance, engine temperature which includes coolant temperature, fuel temperature, oil temperature, etc. Parameter Groups and their numbers (PGN) are listed in SAE J1939 (roughly 300 pages) and defined in SAE J1939/71, a document containing roughly 800 pages filled with parameter group definitions plus suspect parameter numbers (SPN)[21].

Despite common-sense expectations, the SAE J1939/71 standard lacks any explanation what Program Groups (and their numbers) and Suspect Parameters (and their numbers) are.

3.2.8 SAE J1939/73 Application Layer - Diagnostics

J1939/73 defines functions and messages for accessing diagnostic and calibration data. There are several predefined Diagnostic Messages (DM) used for:

- Reading and writing to ECU memory
- Reporting diagnostic information when running
- Identification of lamp status
- Reading and clearing Diagnostic Trouble Codes (DTCs)
- Start/stop broadcast DMs

[21] The definition of Parameter Groups and associated Parameter Group Numbers and Suspect Parameter Numbers is beyond the scope of this book.

3.2.9 SAE J1939/74 Application – Configurable Messaging

J1939/74 describes the message structure for a set of messages that enable the user to determine and announce the parameter placement within a particular message.

3.2.10 SAE J1939/75 Application Layer – Generator Sets and Industrial

J1939/75 describes the parameters and parameter groups predominantly associated with monitoring and control generators and other driven equipment for electric power generation as well as industrial applications.

3.2.11 SAE J1939/81 Network Management

Network Management under J1939 is primarily represented by the Address Claiming Process. While other higher layer protocols based on Controller Area Network (CAN) do not support dynamic node address assignments per default, the SAE J1939 standard provides this ingeniously designed feature to uniquely identify ECUs and their primary function.

J1939-81 provides information about the architecture of an ECU Name and how the ECU claims an addressing using that Name. The Name is a 64 bit (8 bytes) long number giving every ECU a unique identity.

It is somewhat troubling that SAE J1939/81 allows (and describes) a situation during the address claim process where two CAN nodes with identical message IDs can access the bus at the same time. This scenario is, if the CAN standard has any significance at all, not legitimate. SAE J1939/81 recommends a procedure to solve this situation that can only be characterized as peculiar.

J1939 Message Format

The main document describing the J1939 message format is SAE J1939/21 – Data Link Layer. J1939/21 defines the use of the CAN data frame (29-bit identifier, Parameter Group Numbers – PGN, etc.) and the transport protocol functions, i.e. a definition of how messages longer than the standard CAN data length (8 bytes) are transmitted in a J1939 bus network[22].

Without further ado, the J1939/21 document, assuming that the reader has already studied all applicable publications and knows what Parameter Groups are[23], jumps immediately into the detailed architecture of Parameter Group Numbers. It is like explaining the function of an automobile by starting with the details of the gas injection system. The document is poorly written and lacks any structure.

For instance, the mere fact that some J1939 messages may be longer than the standard CAN message of 8 bytes is mentioned in numerous chapters without offering any substantial details on how the messages are packetized. In the same spirit, topics like, for instance, PDU Format are explained repeatedly, spanning over several chapters.

Most irritating, however, is the fact that the SAE engineers who created the standard are not familiar with the unit of time, "ms" or "msec" (milli-seconds). Instead they used mS, which is officially milli-Siemens (electric conductance, equal to inverse Ohm - Ω).

[22] This statement is based on SAE claims. In all consequence the document does not offer any detailed description that would enable the reader to understand the packaging of messages longer than 8 bytes.
[23] As a matter of fact, the document does not provide any explanation of what Parameter Groups are.

4.1 Extended Message Identifier

J1939 uses only the CAN 29-Bit message identifier (In fact, the CAN standard was extended from 11 to 29 bit per request by the SAE in order to support J1939). J1939 uses the identifier, among other features, to identify the source and, in some cases, the destination of data on the bus.

The CAN standard in itself does not support node (ECU) addresses, only message IDs. The sender of message is not concerned with who receives the data, while the receiver does not know who sent the data. The receiver knows the meaning of the data by looking at the message ID. The SAE J1939 Standard ingeniously uses the 29-Bit message identifier to include a source address and, depending on the operating mode, the destination address.

The 29 bit message identifier consists of the regular 11 bit base identifier and an 18 bit identifier extension. The distinction between CAN base frame format and CAN extended frame format is accomplished by using the IDE bit inside the Control Field. A low (dominant) IDE bit indicates an 11 bit message identifier, a high (recessive) IDE bit indicates a 29 bit identifier.

An 11 bit identifier (standard format) allows a total of 2^{11} (= 2048) different messages. A 29 bit identifier (extended format) allows a total of 2^{29} (= 536+ million) messages.

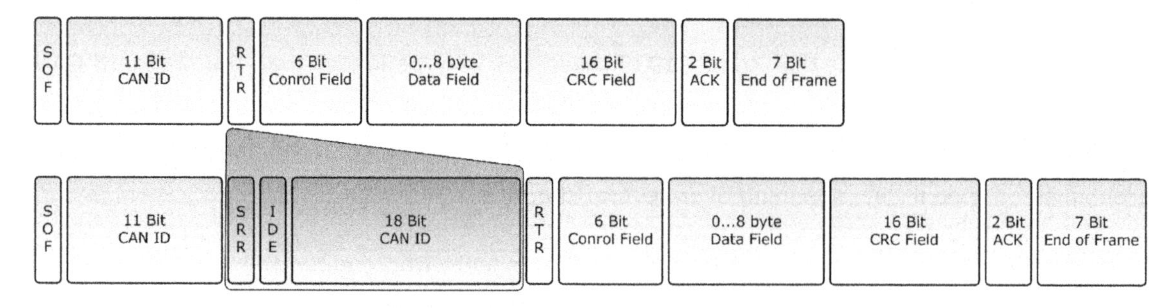

Picture 4.1.1 Extension from 11-Bit to 29-Bit CAN Identifier

Picture 4.1.1 shows a comparison between a standard CAN data frame with an 11-Bit identifier and a CAN data frame in extended format (29-Bit identifier). Both frames contain an Identifier Extension Bit (IDE)[24], which is at low level for the standard frame and at high for the extended data frame. CAN controllers must be designed in a way that they check the IDE in order to distinguish between the two possible frame formats.

While the standard J1939 protocol utilizes the CAN 29-Bit identifier, it nevertheless does allow the sharing of the network with devices that support the standard CAN 11-Bit message identifier[25]. As a matter of fact, this feature is not specific to J1939, but the CAN protocol. SAE J1939 does, however, not provide any further definition on the use of the 11-Bit identifier.

Hear Ye! Hear Ye!

Both formats, Standard (11 bit message ID) and Extended (29 bit message ID), may co-exist on the same CAN bus. During bus arbitration the standard 11 bit message ID frame will always have higher priority than the extended 29 bit message ID frame with identical 11 bit base identifier and thus gain bus access.[26]

In case where an application needs to use both formats, it is recommended to assign a specific priority to all 11-Bit ID messages, that is not being used by the 29-Bit ID messages. This will prevent that any 11-Bit message overrides a 29-Bit message and (in the spirit of the overcautious SAE requirements) eliminate any potential for transmission failures.

The Extended Format has some trade-offs: The bus latency time is longer (minimum 20 bit-times), messages in extended format require more bandwidth (about 20 %), and the error detection performance is reduced (because the chosen polynomial for the 15-bit CRC is optimized for frame length up to 112 bits).

---

[24] The IDE in an 11-Bit standard frame is embedded in the Control Field.

[25] The MilCAN protocol as introduced in chapter *Milcan* takes advantage of this feature.

[26] SAE J1939/21 refers to this scenario as "Incorrect bus arbitration", because the source address in the 11-Bit standard frame might have a higher relative priority. First of all, the source address is not connected to the priority and, secondly, a message with 11-Bit identifier does NOT contain a source address.

4.2 Protocol Data Unit

SAE J1939/21 defines the term Protocol Data Unit as a CAN 29-Bit ID message stripped by the CAN control fields such as SOF (Start of Frame), RTR (Remote Transmission Request), the Control field containing the Data Length Code (DLC), the checksum field, the acknowledgement field (ACK) and the End of Frame (EOF) field. What remains is the information to address the data, i.e. the 29-Bit message ID and the actual data (0 to 8 bytes).

The definition of the Protocol Data Unit is of no further significance for the J1939 protocol, but it was necessary to name the most important sections inside the J1939 29-Bit message ID. Its abbreviation (PDU) and especially abbreviations of the abbreviation (PDU Format = PF, PDU Specific = PS, etc.) are being used extensively throughout the SAE J1939 standards collection.

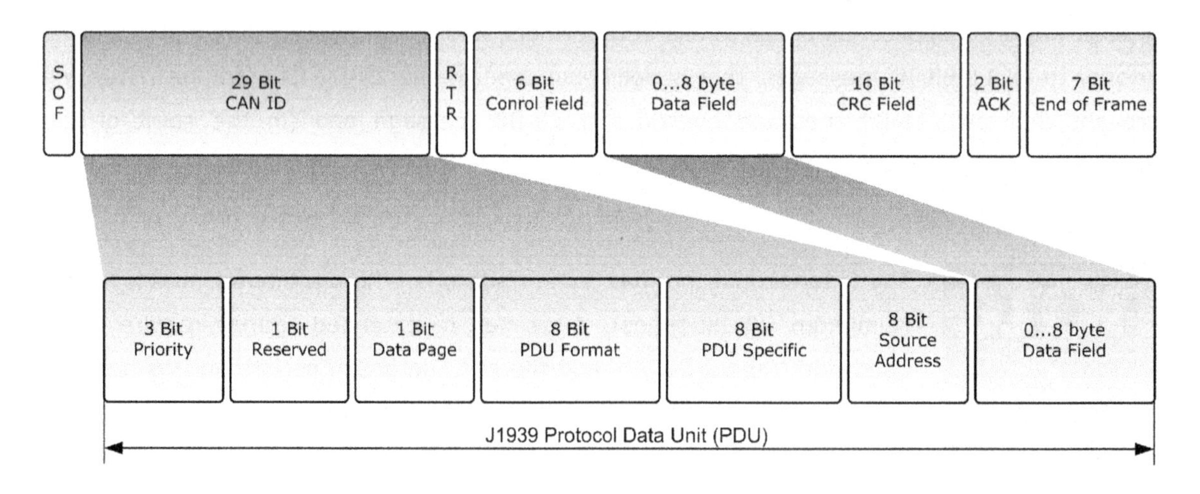

Picture 4.2.1 J1939 Protocol Data Unit (PDU)

The following picture demonstrates the use of the 29-Bit message ID[27].

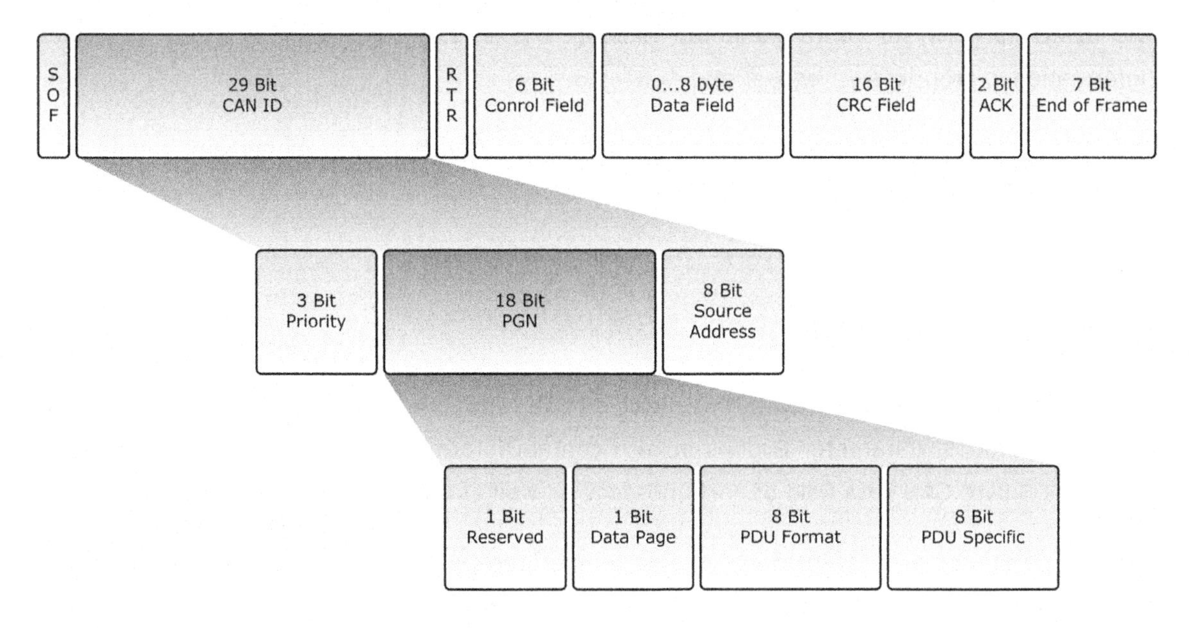

Picture 4.2.2 J1939 Message Format

The parameters embedded in the 29-Bit message identifier are divided into three sections, the priority field, PGN (Parameter Group Number)[28] and the 8 bit source address.

Priority

The first three bits in the identifier represent the priority during the arbitration process, thus providing eight priority levels. In compliance with the CAN standard a value of 0 (000) has the highest priority; a value of 8 (111) has the lowest priority.

[27] For further detailed technical information on the CAN message frame architecture please refer to "A Comprehensible Guide to Controller Area Network" by the same author and publisher. http://www.copperhillmedia.com.

[28] For a peer-to-peer communication the PDU Specific field is used as destination address and, in this case, is not part of the PGN. Also, the actual PGN is internally extended to 24 bits. See also chapter *Parameter Group Numbers*.

High priority messages would be assigned to time critical data such as, for instance, torque control data from the transmission to the engine. Lower level priorities are suitable for non-time-critical data such as, for instance, engine configuration data.

The default priority for control oriented messages is 3; the default for all other messages (informational, proprietary, request and ACK) is 6.

Each PGN is assigned a recommended priority, but for the purposes of network tuning, the priority may be programmable for each message type.

Parameter Group Number (PGN)

Parameter Group Numbers point to vehicle signals and parameters which are defined as Parameter Groups. Parameter Groups in turn contain information on parameter assignments within the 8 byte CAN data field of each message as well as repetition rate and priority.

Parameters groups are, for instance, engine temperature which includes coolant temperature, fuel temperature, oil temperature, etc. Parameter Groups and their numbers (PGN) are listed in SAE J1939 (roughly 300 pages) and defined in SAE J1939/71, a document containing roughly 800 pages filled with parameter group definitions plus suspect parameter numbers (SPN)[29].

The actual PGN is internally extended to 24 bits (See also chapter *Parameter Group Numbers).*

The sections of a Parameter Group Number (PGN) are R (Reserved), Data Page (DP), PDU Format (PF) and PDU Specific (PS).

The structure of the PGN allows a total of 8672 different Parameter Groups per page.

Source Address

The last 8 bits of the 29-Bit message identifier contain the source address, i.e. the address of the transmitting ECU (node). There is a total of 253 addresses available and every address must be unique within the network, i.e. ECUs cannot share addresses. PGNs, however, are independent of the source address, meaning every ECU is allowed to transmit any message.

[29] The definition of Parameter Groups and associated Parameter Group Numbers and Suspect Parameter Numbers is beyond the scope of this book.

4.3 Communication Methods

SAE J1939 provides three communication methods, each serving a specific purpose.

1. Destination Specific Communications:

Destination specific communications use PDU1 (PF values 0 to 239), but also the global destination address 255. There are cases where this method will require the utilization of destination specific Parameter Group Numbers, for instance, in the case of more than one engine. A torque message, for example, must be sent only to the desired engine and not to both.[30]

2. Broadcast Communications

Broadcast communications use PDU2 (PF values 240 to 255) and, as the name implies, they can include:

- Sending a message from a single or multiple sources to a single destination.
- Sending a message from a single or multiple sources to multiple destinations.

3. Proprietary Communications

Proprietary communications use either PDU1 or PDU2 and, as the name implies, they are useful in case where standard communications are not practical[31].

The use of PDU1 or PDU2 indicates that there may be:

- **Broadcast Proprietary Communications**

and

- **Destination Specific Proprietary Communications**

A Parameter Group Number (PGN) has been assigned for both proprietary communication types.

[30] The SAE J1939 document uses the expression "...the message must be directed to one or another specific destination and not both."
[31] The SAE J1939 document uses the expression "Where it is important to communicate proprietary information." - yet another example of pointless redundancy.

The reception and processing of received messages are explained in detail in SAE J1939/21 (Data Link Layer) and J1939/7x (Application Layer).

In general a received message is handled according to the communication method[32]:

- **Destination Specific Request or Command**

  Each receiving ECU must determine whether the incoming destination address matches its own address and if yes, it must process the message and respond accordingly[33].

- **Global Request**

  Each ECU in the network, even the sender of the request, must process the message and respond if the requested data is available.

- **Broadcast**

  Each ECU must determine individually whether or not the message is relevant.

4.4 Parameter Group Numbers

SAE J1939 is a very ingeniously designed protocol that takes a resourceful advantage of the CAN 29-Bit message identifier. Rather than relying on a myriad of protocol functions, SAE J1939 uses predefined parameter tables, which keeps the actual protocol on a comprehensible level. However, these parameter tables (Parameter Groups) are also the biggest stumbling block when it comes to implementing the protocol into an embedded solution (ECU).

Parameters groups are, for instance, engine temperature which includes coolant temperature, fuel temperature, oil temperature, etc. Parameter Groups and their numbers (PGN) are listed in SAE J1939 (roughly 300 pages) and defined in SAE J1939/71, a document containing

[32] The SAE J1939 document uses the sentence "Several general observations can be made however regarding received messages." which not only lacks the proper grammar, but it also raises the question whether SAE J1939 is a research project (hence "observations") or a Standard where network specifics are outlined.
[33] SAE J1939 uses the vague wording "...provide some type of acknowledgement."

roughly 800 pages filled with parameter group definitions plus suspect parameter numbers (SPN)[34].

The structure of the PGN allows a total of 8672 different Parameter Groups per page (2 pages are available).

The SAE offers a "J1939 Companion Spreadsheet", an Excel file which is supposed to supplement the J1939 standards collection. It consists of parameters and parameter groups contained in SAE J1939 and SAE J1939/71. The spreadsheet is available for purchase through the SAE web site for a price of US$125.00 for non-members.

However, the document is also protected by SAE copyrights which specifically prohibit the copying of any of the information contained in the file, for instance, to use it in a J1939 application. A copyright can be obtained through the SAE Copyright Administrator for a mere US$30,000 per year.

4.4.1 Parameter Groups (PG)

Parameters groups are, for instance, engine temperature which includes coolant temperature, fuel temperature, oil temperature, etc. The Parameter Groups (PG) architecture and Parameter Group Numbers (PGN) are described in SAE J1939/21[35] and listed in SAE J1939 (roughly 300 pages) and defined in SAE J1939/71, a document containing roughly 800 pages filled with parameter group definitions[36]. In addition, it is possible to use manufacturer-specific parameter groups.

Parameter Groups contain information on parameter assignments within the 8 byte CAN data field of each message as well as repetition rate and priority.

[34] The assignment of Parameter Groups and associated Parameter Group Numbers and Suspect Parameter Numbers is beyond the scope of this book.

[35] Parts of the standard describing the Parameter Group Numbers and how they are embedded in a CAN message are either based on a misunderstanding of the CAN standard or documenting deficiencies.

[36] The assignment of Parameter Groups, Parameter Group Numbers and Suspect Parameter Numbers is beyond the scope of this book.

The following is an example of a parameter group definition as listed in SAE J1939/71:

PGN 65262	**Engine Temperature**
Transmission Rate	1 sec
Data Length	8 bytes
Data Page	0
PDU Format (PF)	254
PDU Specific (PS)	238
Default Priority	6
PG Number	65262 ($FEEE_{hex}$)

Description of Data			**SPN**
Byte	1	Engine Coolant Temperature	110
	2	Fuel Temperature	174
	3, 4	Engine Oil Temperature	175
	5, 6	Turbocharger Oil Temperature	176
	7	Engine Intercooler Temperature	52
	8	Engine Intercooler Thermostat Opening	1134

4.4.2 Suspect Parameter Number (SPN)

A Suspect Parameter Number[37] (SPN) is a number assigned by the SAE to a specific parameter within a parameter group. It describes the parameter in detail by providing the following information:

- Data Length in bytes
- Data Type
- Resolution
- Offset
- Range
- Reference Tag (Label)

[37] The term "Suspect" is apparently meant in regards to "assume, presume, expect".

SPNs that share common characteristics are grouped into Parameter Groups (PG) and they will be transmitted throughout the network using the Parameter Group Number (PGN).

To follow up on the previous example (PGN 65262), the parameter Engine Coolant Temperature is described by SPN 110 in the following way:

SPN 110	**Engine Coolant Temperature**
Temperature of liquid engine cooling system	
Data Length	1 Byte
Resolution	1 deg C / Bit
Offset	-40 deg C
Data Range	-40 to 210 deg C
Type	Measured
Reference	PGN 65262

The following picture demonstrates the use of Suspect Parameter Numbers, Parameter Groups and Parameter Group Numbers.

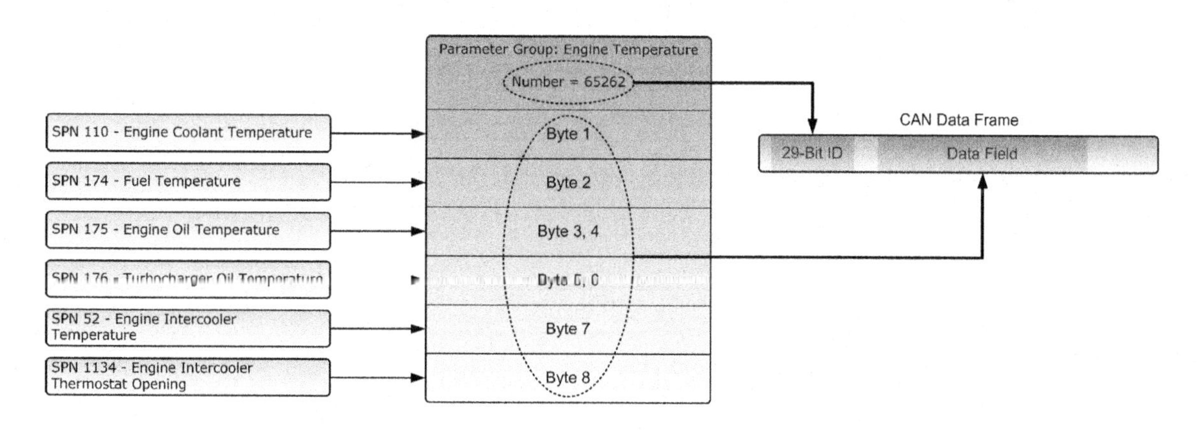

Picture 4.4.2.1 SPNs and PGNs

The Suspect Parameter Numbers in this example point to vehicle data associated with the Engine Temperature. Note that all SPNs have been selected logically to fit into the Parameter Group *Engine Temperature*.

The Parameter Group *Engine Temperature* has been assigned the Parameter Group Number (PGN) 65262. It is in fact the Parameter Group Number (embedded in the 29-Bit message ID) plus the actual data (embedded in the CAN data field) that is being transmitted into the CAN bus.

It is a requirement that all nodes sending or receiving the PGN 65262 know the structure of the PGN as well as all associated SPNs.

4.4.3 Parameter Group Number Architecture

The PGN uniquely identifies the Parameter Group (PG) that is being transmitted in the message. Each PG (a grouping of specific parameters) has a definition that includes the assignment of each parameter within the 8-byte data field (size in bytes, location of LSB), and the transmission rate and priority of the message. The structure of a PGN permits a total of up to 8672 different parameter groups to be defined.

When an ECU receives a message, it uses the PGN in the identifier to recognize the type of data that was sent in the message.

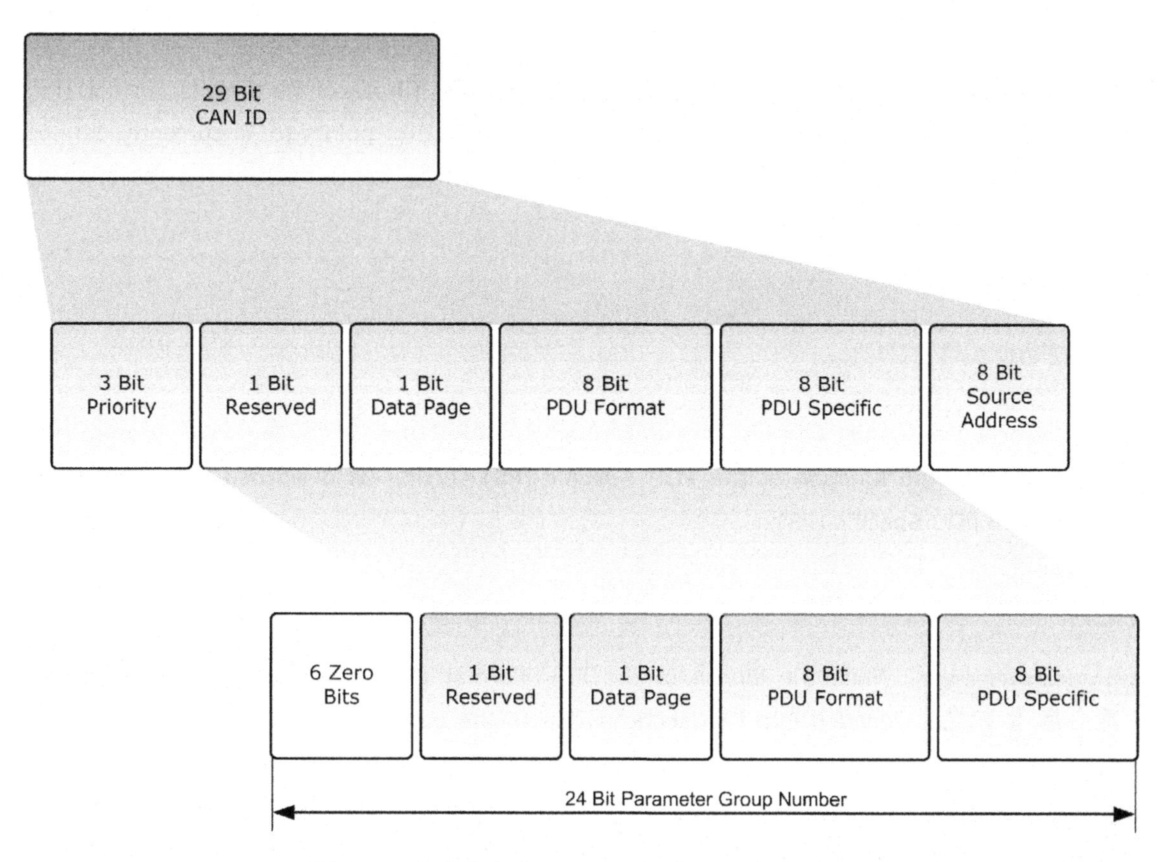

Picture 4.4.3.1 Parameter Group Number

For internal purposes the parameter group number is extended to 24 bits = 3 bytes (See more detailed information in chapter *Parameter Group Number Compilation*).

The following explains the individual sections within the parameter group number section of the 29-Bit message identifier.

R

According to the SAE J1939 document, the R bit is reserved for future purposes and should always be set to 0 when transmitting messages. SAE J1939/21, however, calls this bit Extended Data Page (EDP). It also states that J1939 messages should always set the bit to zero on transmit, but later refers to EDPs who are set to one (ISO 15765-3).

DP – Data Page

The DP bit works as a page selector for the following PDU (Protocol Data Unit) Format (PF) field. Currently this bit is at 0, pointing to page 0, which in turn points to all currently defined messages. Page 1 will be used to provide extended capacity for the future, i.e. as soon as page 0 has reached its capacity.

PDU Format (PF)

PDU (Protocol Data Unit) Formats are described in SAE J1939/21[38]. The PDU Format (PF) basically defines the function of the PDU Specific (PS) section as described in the following paragraph on PDU Specific (PS).

While the length of the PDU Format (PF) is officially 8 bits, it is also divided into two sections due to the construction of the 29-Bit message identifier (11-Bit ID + SRR + IDE + 18-Bit ID). The SRR and IDE bits are entirely defined by the CAN standard 2.0B and thus are not described or modified by the SAE J1939 standard. For the same reason, some documentation may refer to a PF length of 10 bits.

PDU Specific (PS)

PDU Specific means that its content is interpreted according to the information in the PDU Format (PF). A PF value between 0 and 239 (PDU1) indicates a destination address in PS (peer-to-peer communication). A PF value between 240 and 255 (PDU2, broadcast message) indicates a Group Extension (GE) to the PDU Format (PF). The GE is used to increase the number of available messages to be broadcasted throughout the network.

[38] The SAE J1939 Standard refers to *SAE J1939/21 Section 3.3* which does not exist. In turn SAE J1939/21 repeats a large amount of information already present in SAE J1939, however, in a different form.

	PDU Format	PDU Specific	Communication Mode
PDU1 Format	0 - 239 0_{hex} - EF_{hex}	Destination Address	Peer-to-Peer
PDU2 Format	240 - 255 $F0_{hex}$ - FF_{hex}	Group Extension	Broadcasting

Table 4.4.3.1 PDU Format and PDU Specific

While the destination address is defined to address a specific node (ECU) in the network, it can also be used to address all ECUs at the same time. A destination address (DA) of 255 is called a Global Destination Address and it requires all nodes to listen and, if required, to respond. The SAE J1939/21 standard defines the Global Destination Address in one short sentence that can be easily overlooked.

Messages according to SAE J1939 are mostly broadcasted using the PDU2 format (Group Extension). Such messages cannot be transmitted to a specific destination.

The following picture shows the mapping of a J1939 message into the extended CAN 29-Bit message frame:

CAN Extended Frame Architecture

SOF	11 Bit CAN ID	SRR	IDE	18 Bit CAN ID	RTR	...

SOF	3 Bit Priority	R	DP	6 Bit (MSB) PDU Format	SRR	IDE	2 Bit PF Cont.	8 Bit PDU Specific (PS)	8 Bit Source Address	RTR	...

J1939 Frame Architecture

Picture 4.4.3.2 J1939 Message Mapping

Note the high-lighted portion in the J1939 Frame Architecture that points to PDU Format Field (PF). The PF is divided into two sections, separated by the CAN SRR and IDE bit. The SRR and IDE bits are entirely defined by the CAN standard 2.0B and thus are not described or modified by the SAE J1939 standard. For the same reason, some documentation may refer to a PF length of 10 bits.

4.4.5 Parameter Group Number Range

With the definition of PDU Format (PF) and PDU Specific (PS) – as shown below - J1939 supports a total of 8672 Parameter Group numbers.

	PDU Format	**PDU Specific**	**Communication Mode**
PDU1 Format	0 – 239 0_{hex} - EF_{hex}	Destination Address	Peer-to-Peer
PDU2 Format	240 – 255 $F0_{hex}$ - FF_{hex}	Group Extension	Broadcasting

Table 4.4.5.1 PDU Format and PDU Specific

The Parameter Group Number range is divided into two sections:

1. Specific PGNs for peer-to-peer communication (PDU1 Format)

   Range: 00_{hex} - EF_{hex} (not including PDU Specific)

   Number of PGNs: 240

2. Generic PGNs for message broadcasting (PDU2 Format)

   Range: $F000_{hex}$ – $FFFF_{hex}$ (including PDU Specific)

   Number of PGNs: 4096

Considering the Data Page (DP) bit, the total number of PGNs is (240 + 4096) * 2 = 8672.

As a reminder:

The DP bit works as a page selector for the following PDU (Protocol Data Unit) Format (PF) field. Currently this bit is at 0, pointing to page 0, which in turn points to all currently defined messages. Page 1 will be used to provide extended capacity for the future, i.e. as soon as page 0 has reached its capacity.

The following shows a Parameter Group Number map:

DP	PGN Range (hex)	Number of PGNs	SAE or Manufacturer Assigned	Communication
0	000000 - 00EE00	239	SAE	PDU1 = Peer-to-Peer
0	00EF00	1	MF	PDU1 = Peer-to-Peer
0	00F000 - 00FEFF	3840	SAE	PDU2 = Broadcast
0	00FF00 - 00FFFF	256	MF	PDU2 = Broadcast
1	010000 - 01EE00	239	SAE	PDU1 = Peer-to-Peer
1	01EF00	1	MF	PDU1 = Peer-to-Peer
1	01F000 - 01FEFF	3840	SAE	PDU2 = Broadcast
1	01FF00 - 01FFFF	256	MF	PDU2 = Broadcast

Table 4.4.5.1 Parameter Group Number Range

The current range of Parameter Group Numbers as defined in SAE J1939/71 is from PGN 0 (Torque/Speed Control) to PGN 65279 (Water in Fuel Indicator). This range is not a real representation of the total number of PGNs, since there a gaps between PGN definitions. The same is true for SPNs, who range from SPN 16 (Engine Fuel Filter) to SPN 4096 (XBR Urgency).

4.4.6 Parameter Group Number Compilation

For internal purposes the parameter group number is extended to 24 bits = 3 bytes where the most significant 6 bits are always set to zero. This process must be accomplished by each ECU individually; this procedure is not part of the CAN standard.

In order to compile the exact Parameter Group Number (PGN) it is yet again necessary to consider the two alternatives, PDU1 (Peer-to-Peer communication) and PDU2 (Broadcast communication) format.

PDU1 – Peer-to-Peer Communication

The following picture demonstrates how a Parameter Group Number is assembled in PDU1 format. The PDU Format is less than 240 and thus the least significant 8 bits are set to zero. PDU Specific (PS) is handled as the destination address.

While the destination address is defined to address a specific node (ECU) in the network, it can also be used to address all ECUs at the same time. A destination address (DA) of 255 is called a Global Destination Address[39] and it requires all nodes to listen and, if required, to respond. The SAE J1939/21 standard defines the Global Destination Address in one short sentence that can be easily overlooked.

[39] No acronym offered by SAE.

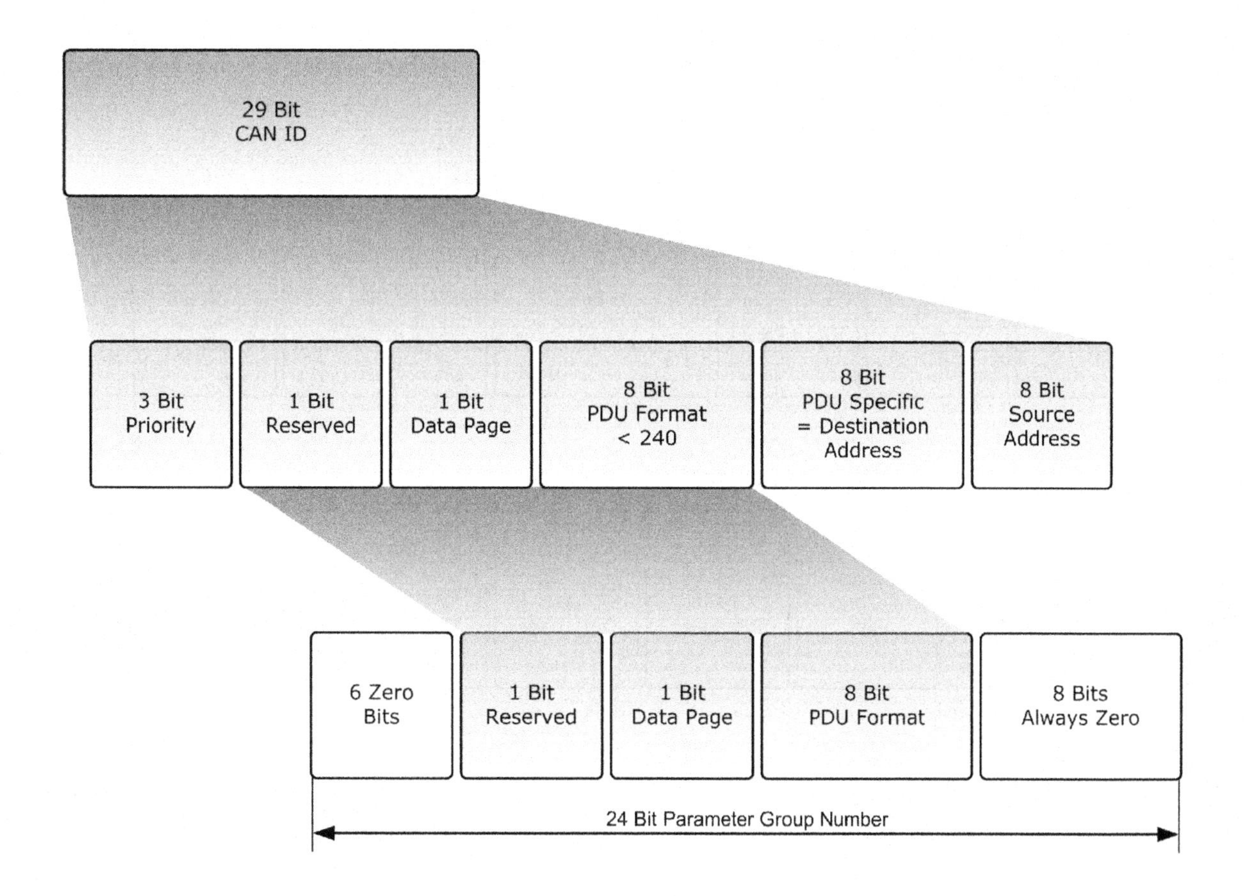

Picture 4.4.6.1 Parameter Group Number – PDU1 Format

PDU2 – Broadcast Communication

The following picture demonstrates how a Parameter Group Number is assembled in PDU2 format. The PDU Format is equal to or greater than 240 and thus PDU Specific (PS) is handled as a group extension (GE).

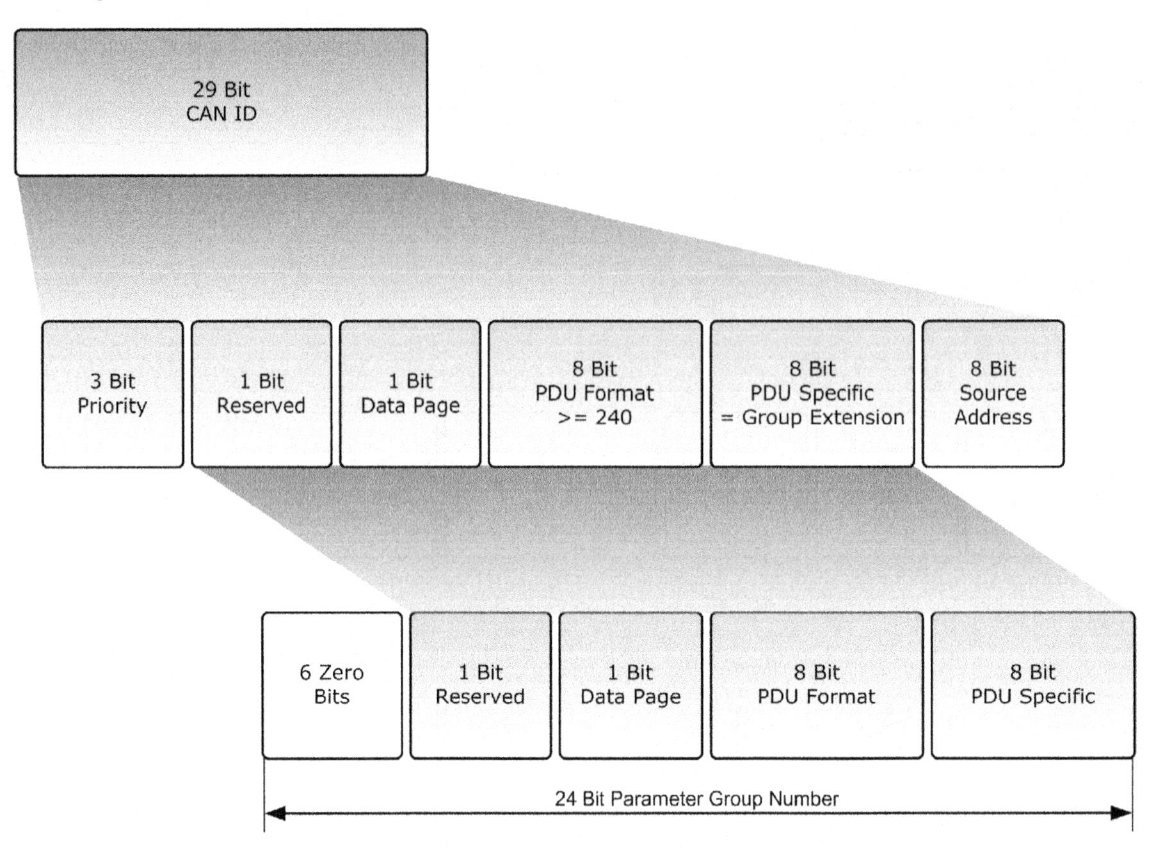

Picture 4.4.6.2 Parameter Group Number – PDU2 Format

4.4.7 Proprietary Parameter Group Numbers

As the name Proprietary Parameter Group Numbers[40] implies, the SAE J1939 standard supports Parameter Groups in PDU1 and PDU2 Format that can be assigned by manufacturers for their specific needs, which includes the use of the data field in the message. The only constraint is that, per SAE request, an "excessive" utilization of the network (e.g. 2 percent or more) should be avoided. This requirement is, however, not based on any technical restrictions by the CAN/J1939 network.

The following table shows the Parameter Group Number Range where the Proprietary Parameter Group Numbers are highlighted.

[40] In this case the SAE J1939/21 standard is not using an acronym such as "PPGN".

DP	PGN Range (hex)	Number of PGNs	SAE or Manufacturer Assigned	Communication
0	000000 - 00EE00	239	SAE	PDU1 = Peer-to-Peer
0	00EF00	1	MF	PDU1 = Peer-to-Peer
0	00F000 - 00FEFF	3840	SAE	PDU2 = Broadcast
0	00FF00 - 00FFFF	256	MF	PDU2 = Broadcast
1	010000 - 01EE00	239	SAE	PDU1 = Peer-to-Peer
1	01EF00	1	MF	PDU1 = Peer-to-Peer
1	01F000 - 01FEFF	3840	SAE	PDU2 = Broadcast
1	01FF00 - 01FFFF	256	MF	PDU2 = Broadcast

Table 4.4.7.1 Parameter Group Number Range

Proprietary Parameter Groups and their numbers are designed using the exact same structure as Parameter Group and their numbers defined by the SAE.

Parameter Group Name	**Proprietary A**
Parameter Group Number	61184 ($00EF00_{hex}$)
Definition	Proprietary PG using the PDU1 Format for Peer-to-Peer communication.
Transmission Rate	Manufacturer Specific
Data Length	0 – 1785 bytes (multi-packet supported)
Extended Data Page (R)	0
Data Page	0
PDU Format	239
PDU Specific	8 bit Destination Address - Manufacturer Assigned
Default Priority	6
Data Description	Manufacturer Specific

Parameter Group Name	**Proprietary A2**
Parameter Group Number	126720 ($01EF00_{hex}$)
Definition	Proprietary PG using the PDU1 Format for Peer-to-Peer communication.
Transmission Rate	Manufacturer Specific
Data Length	0 – 1785 bytes (multi-packet supported)
Extended Data Page (R)	0
Data Page	1
PDU Format	239
PDU Specific	8 bit Destination Address – Manufacturer Assigned
Default Priority	6
Data Description	Manufacturer Specific

Parameter Group Name	**Proprietary B**
Parameter Group Number	65280 - 65535 ($00FF00_{hex}$ – $00FFFF_{hex}$)
Definition	Proprietary PG using the PDU2 Format for Broadcast communication.
Transmission Rate	Manufacturer Specific
Data Length	0 – 1785 bytes (multi-packet supported)
Extended Data Page (R)	0
Data Page	0
PDU Format	255
PDU Specific	Group Extension – Manufacturer Assigned
Default Priority	6
Data Description	Manufacturer Specific

For reasons unknown the SAE J1939/21 standard does not define a Proprietary B2 Parameter Group Name, which would be the same as Proprietary B, but with Data Page = 1. Setting Data Page = 1 is used, for instance, to define Proprietary A2. In all consequence there is no reason not to define Proprietary B2.

The use of Proprietary Parameter Group Numbers does, however, come with a downside, that is briefly mentioned by SAE J1939/21 in the Proprietary B PGN definition. In fact, the problem does apply to all Proprietary Parameter Group Numbers.

The potential problem comes with the possibility that two or more manufacturers may use the same Proprietary Parameter Group Number. SAE J1939/21 recommends that receivers of such messages need to differentiate between manufacturers, which can be accomplished by checking the Manufacturer Code in the ECU's NAME field (see also chapter *ECU Name and Addresses*).

4.4.8 Parameter Group Number Assignments

Parameter Group Numbers are assigned to use the PDU 1 format or the PDU 2 format (See also paragraph *J1939 Message Format*). The PDU format cannot be changed once it is assigned to a PGN.

When assigning a Parameter Group Number (PGN) the following criteria should be taken into account:

- Priority
- Update Rate
- Importance of the data to other ECUs
- Data length associated with the Parameter Group

While the use of standardized communications is preferred there may be, nevertheless, unique requirements.

For instance, there may be cases where a number of ECUs in a J1939 network come from the same manufacturer and thus the communication between these ECUs may be highly manufacturer-specific (i.e. proprietary) and of no use for the other ECUs in the network. In such a case the SAE J1939 standard recommends the use of proprietary Parameter Groups.

The choice of broadcast format can be categorized as follows:

PDU 1	The information is of general interest, but requires destination specific addressing.
PDU 2	The information is of general interest, but does not require destination specific addressing.
Proprietary	The information is either strictly proprietary or of no general interest.

PDU 2 should be the preferred broadcast format, while PDU 1 and Proprietary formats should be considered carefully and used only when absolutely necessary.

4.4.9 Data Field Grouping

Considering modern microprocessor technologies, the standard J1939 baud rate of 250 kBit/sec posts no real challenge to a CAN network in regards to performance even at a high bus load. The SAE J1939 Standard, nevertheless, recommends a frugal approach when it comes to assigning data in a CAN message frame plus minimizing the network traffic, while pointing out that the standard provides guidelines rather than dictating strict rules. Yet again, the ground-rule is that a lean and well-thought system design will help avoiding potential problems.

The standard CAN message (regardless of message ID length) includes a data field of flexible length, i.e. between 0 and 8 bytes (See next picture).

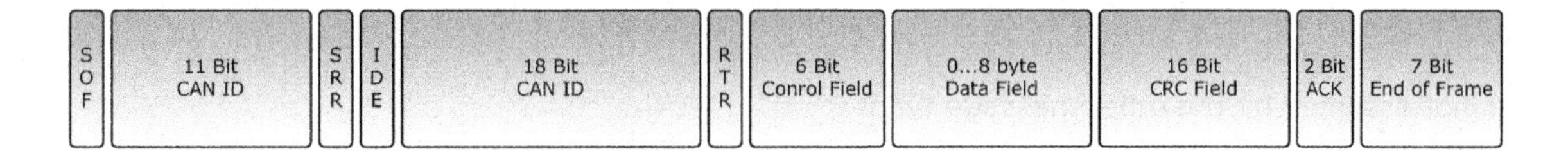

Picture 4.4.9.1 CAN Message Frame

The main objective, according to the SAE J1939 Standard, should be to fill the data field to its full extend of 8 bytes whenever possible.

In order to accomplish the lowest possible network traffic and prevent the under-utilization of the data field the standard recommends to group parameters as follows:

1. By common sub-system (preferably data from one specific ECU)
2. With similar update rates
3. By function (oil, coolant, fuel, etc.)

For messages assigned with a slow update rate it is nevertheless acceptable that not all of the parameters in a Parameter Group come from the same ECU.

4.4.10 Message Types

SAE J1939/21 currently documents a total of five message types[41]. They are:

- Command
- Request
- Broadcast/Response
- Acknowledgement
- Group Functions

[41] SAE J1939/21 also supports special message types, Request2 and Transmit, which are beyond the scope of this comprehensible guide.

The Request, Acknowledgement and Group Functions message type are assigned specific PGNs. Command and Broadcast/Response message type may be associated with any PGN that is not assigned by the other message types.

4.4.10.1 Command

The definition of a Command message type serves only the purpose of having a description of a PGN that is interpreted as a command. It is not associated to a specific PGN or PGN range. The SAE J1939/21 standard wastes only three inflated sentences on the description of the Command message type.

The Command message type is nothing else but an ordinary PGN supporting both PDU formats, PDU1 for Peer-to-Peer communication and PDU2 for Broadcast communication.

4.4.10.2 Request

As the name implies, the Request message type is being used to request data globally or from a specified destination.

The SAE J1939/21 fails to mention that the Request message type supports only the PDU1 format (Peer-to-Peer communication, see also the next note). The response, however, can be PDU1 or PDU2, even multi-packet.

The Request message type is associated with a specific PGN as described below.

Parameter Group Name	**Request**
Parameter Group Number	59904 ($00EA00_{hex}$)
Definition	Requests a Parameter Group from a single device or all devices in the network.
Transmission Rate	User defined (no more 2 to 3 times a second is recommended)
Data Length	3 bytes (CAN DLC = 3)
Extended Data Page (R)	0
Data Page	0
PDU Format	234
PDU Specific	Destination Address (Global or Specific – See following note)
Default Priority	6
Data Description	Byte 1, 2, 3 = Requested Parameter Group Number

While the destination address is defined to address a specific node (ECU) in the network, it can also be used to address all ECUs at the same time. A destination address (DA) of 255 is called a Global Destination Address[42] and it requires all nodes to listen and, if required, to respond. The SAE J1939/21 standard defines the Global Destination Address in one short sentence that can be easily overlooked.

The following table demonstrates the use of fields of a Request message type.

Message Type	**PGN**	**PS (DA)**	**SA**	**Data 1**	**Data 2**	**Data 3**
Global Request	59904	255	Requester	PGN (LSB)	PGN	PGN (MSB)
Specific Request	59904	Responder	Requester	PGN (LSB)	PGN	PGN (MSB)

Table 4.4.10.2.1 Request Message Type - Use of Fields

[42] No acronym offered by SAE.

A node, receiving a request, is required to respond in any event, even when it means to send a send a negative acknowledgement (NACK) when the requested PGN does not exist (See also chapter *Acknowledgement*). A global request will not trigger a NACK when a particular PGN is not supported.

4.4.10.3 Broadcast/Response

The definition of a Broadcast/Response message type serves only the purpose of having a description of a PGN that is interpreted as a data message. It is not associated to a specific PGN or PGN range. The SAE J1939/21 standard wastes only one short sentence on the description of the Broadcast/Response message type.
The Broadcast/Response message type is nothing else but an ordinary PGN supporting both PDU formats, PDU1 for Peer-to-Peer communication and PDU2 for Broadcast communication. It may be an unsolicited broadcast of data or it can be a response to a Command or Request.

4.4.10.4 Acknowledgement

The SAE J1939/21 standard refers to two forms of acknowledgements, one embedded in the CAN frame per standard and one supported by the Acknowledgement message type. The description of the standard CAN acknowledgment field is, however, incorrect and shall not be repeated here. The standard CAN ACK field does not confirm the reception of a data frame. The acknowledgement field serves as a confirmation of a successful CRC (checksum) check by the receiving nodes in the network.

The J1939 Acknowledgement message type, as its name implies, is a response to a command or a request. The SAE J1939/21 fails to mention that the Acknowledgement message type supports only the PDU1 format (Peer-to-Peer communication).

The Acknowledgement message type is associated with a specific PGN as described below.

Parameter Group Name	**Acknowledgement**
Parameter Group Number	59392 (00E800$_{hex}$)
Definition	Provides handshake between transmitting and responding nodes.
Transmission Rate	Upon reception of a command or request.
Data Length	8 bytes (as described in the following)
Extended Data Page (R)	0
Data Page	0
PDU Format	232
PDU Specific	Destination Address (Global = 255)[43]
Default Priority	6
Data Description	Bytes 1…8 = Positive Acknowledgement, Negative Acknowledgement, Access Denied or Cannot Respond (See following description).

Positive Acknowledgement (ACK)	**Byte**	**Value**
	1	Control byte = 0
	2	Group Function Value (if applicable)
	3-4	Reserved; should be filled with FF$_{hex}$.
	5	Address Acknowledged
	6	PGN of requested data - LSB
	7	PGN of requested data
	8	PGN of requested data – MSB
Negative Acknowledgement (NACK)	1	Control byte = 1
	2	Group Function Value (if applicable)
	3-4	Reserved; should be filled with FF$_{hex}$.
	5	Address NACK
	6-8	PGN of requested data

[43] Setting the destination to 255 makes it possible to filter one CAN identifier for all ACK messages.

Access Denied (PGN supported, but security denied access)	1	Control byte = 2
	2	Group Function Value (if applicable)
	3-4	Reserved; should be filled with FF_{hex}.
	5	Address Access Denied
	6-8	PGN of requested data
Cannot Respond (PGN is supported, but ECU cannot respond; request data later)	1	Control byte = 3
	2	Group Function Value (if applicable)
	3-4	Reserved; should be filled with FF_{hex}.
	5	Address Busy
	6-8	PGN of requested data

4.4.10.5 Group Functions

The Group Functions message type is used for special function groups, especially proprietary functions, but also network management and multi-packet transport functions[44].

Parameter Group Name	**Proprietary A**
Parameter Group Number	61184 ($00EF00_{hex}$)
Definition	Proprietary PG using the PDU1 Format for Peer-to-Peer communication.
Transmission Rate	Manufacturer Specific
Data Length	0 – 1785 bytes (multi-packet supported)
Extended Data Page (R)	0
Data Page	0
PDU Format	239
PDU Specific	8 bit Destination Address – Manufacturer Assigned
Default Priority	6
Data Description	Manufacturer Specific

[44] SAE J1939/21 does not elaborate on network management or multi-packet transport functions in terms of Group Function message types.

Parameter Group Name	**Proprietary A2**
Parameter Group Number	126720 ($01EF00_{hex}$)
Definition	Proprietary PG using the PDU1 Format for Peer-to-Peer communication.
Transmission Rate	Manufacturer Specific
Data Length	0 – 1785 bytes (multi-packet supported)
Extended Data Page (R)	0
Data Page	1
PDU Format	239
PDU Specific	8 bit Destination Address – Manufacturer Assigned
Default Priority	6
Data Description	Manufacturer Specific

Parameter Group Name	**Proprietary B**
Parameter Group Number	65280 - 65535 ($00FF00_{hex}$ – $00FFFF_{hex}$)
Definition	Proprietary PG using the PDU2 Format for Broadcast communication.
Transmission Rate	Manufacturer Specific
Data Length	0 – 1785 bytes (multi-packet supported)
Extended Data Page (R)	0
Data Page	0
PDU Format	255
PDU Specific	Group Extension – Manufacturer Assigned
Default Priority	6
Data Description	Manufacturer Specific

4.5 Transport Protocol Functions

Even though extremely effective in passenger cars and small industrial applications, CAN alone was not suitable to meet the requirements of truck and bus communications, especially since its communication between devices is limited to only 8 bytes per message. However, it is possible to extend the size of a CAN message by implementing additional software, i.e. so-called higher layer protocols. J1939 is such a higher layer protocol and it supports up to 1785 bytes per message.

In order to support a size of more than 8 bytes the message needs to be packaged into a sequence of 8 byte size messages. Consequently, the receiver of such a multi-packet message must re-assemble the data. Such functions are defined as Transport Protocol (TP) Functions and they are also described in SAE J1939/21. The two major parts of the TP Functions are Message Packaging[45] & Reassembly and Connection Management. In addition, the Transport Protocol Functions handle flow control and handshaking features for destination specific transmissions.

The SAE J1939/21 standard fails to mention that the multi-packet messages are only supported in PDU1 format (Peer-to-Peer communication). In order to broadcast messages they must be addressed to the global destination address (= 255).

4.5.1 Message Packaging and Reassembly

Certain parameter groups may require more than the 8 data bytes supported by the CAN standard. The SAE J1939 standard[46], namely the Transport Protocol Function, supports message lengths up 1785 bytes. In case a program group requires more than 8 data bytes

[45] The SAE Standard uses the term "Packetization", which is not an official English word.

[46] The SAE J1939/21 standard "wastes" less than one page on the subject of message "packetization" and re-assembly.

(9...1785 bytes) and is defined as multi-packet capable, the data will be packaged into a series of CAN data frames.

A regular CAN data frame includes a data field of variable length, 0 to 8 bytes. The information regarding the transmitted data length is found in the DLC (Data Length Code) inside the CAN Control Field (See following pictures).

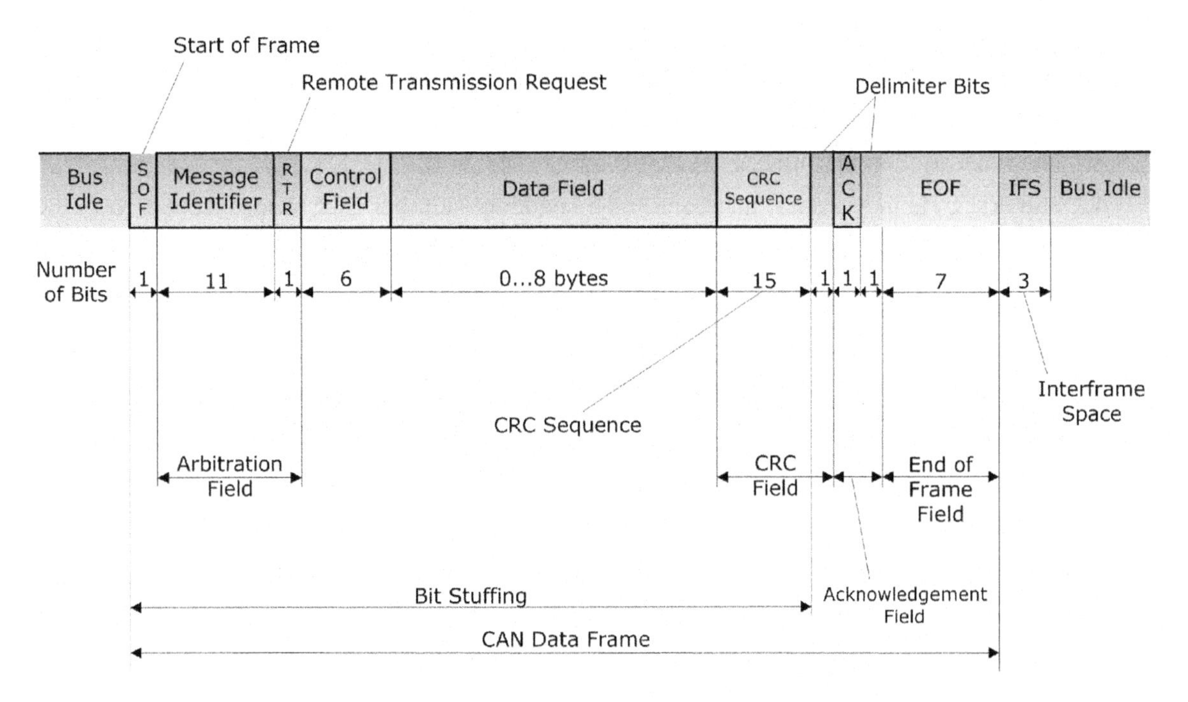

Picture 4.5.1.1 CAN Data Frame Architecture

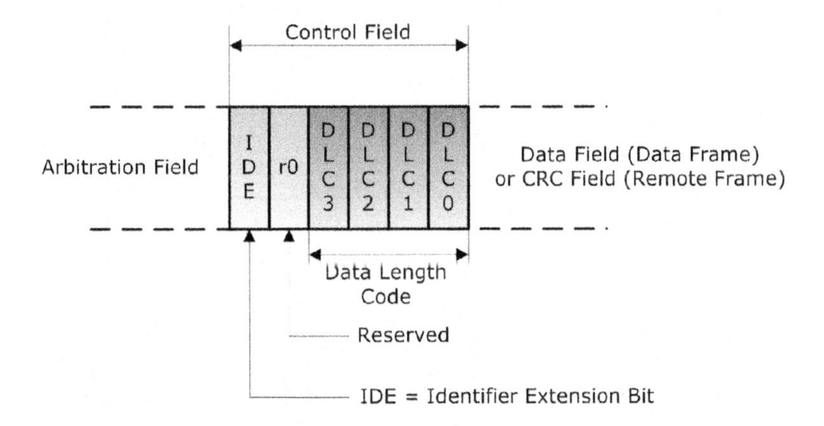

Picture 4.5.1.2 CAN Control Field

In order to package CAN messages into a sequence of up to 1785 messages (as well as to re-assemble the CAN frames into one data package) the J1939 Transport Protocol defines the following:

1. Each multi-packet message is being transmitted by using a dedicated Data Transfer PGN (60160, TP.DT = Transfer Protocol Data Transfer), i.e. all message packets will have the same ID[47].
2. The flow control is managed by another dedicated PGN (60146, TP.CM = Transfer Protocol Communication Management).
3. The message length must always be 8 bytes (DLC = 8).
4. The first byte in the data field contains a sequence number that ranges from 1 to 255.
5. The remaining 7 bytes are filled with the data of the original long (> 8 bytes) message.
6. All unused data bytes in the last package are being set to FF_{hex}.

The actual total message length is defined by the corresponding Parameter Group (See also chapter *Multi-Packet Broadcast*).

The method of using a sequence number plus the remaining seven data bytes yields a total of (255 packages times 7 bytes/package) 1785 bytes per multi-packet message.

The following pictures demonstrate the use of the CAN data field including the sequence number and the packaging of multiple CAN messages.

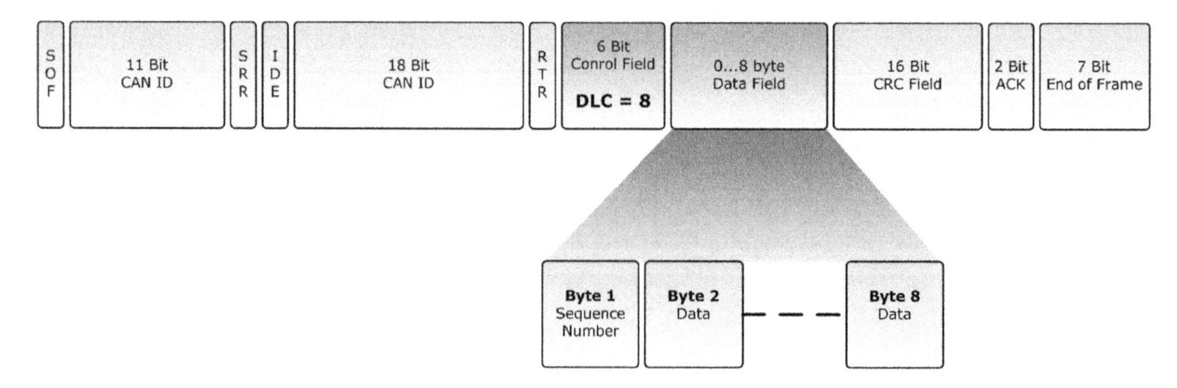

Picture 4.5.1.3 CAN Data Frame with Sequence Number

[47] This important requirement is mentioned only casually at the end of the brief "Packetization" paragraph inside SAE J1939/21.

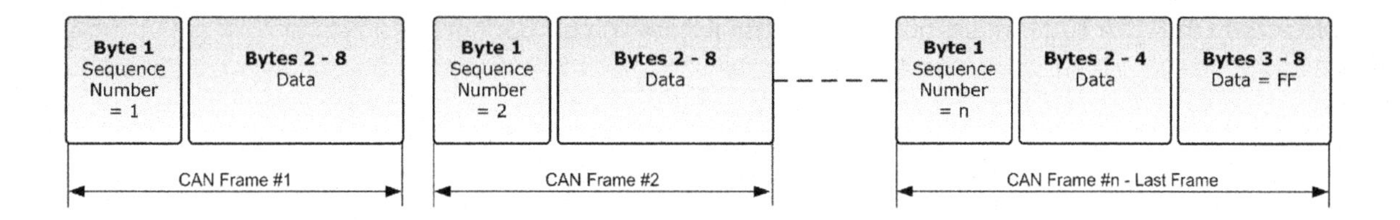

Picture 4.5.1.4 Example of Multi-Package Sequence

The data packages are re-assembled in the order of their sequence number and the final data package can then be passed to the application layer.

4.5.2 Connection Management

The J1939 Connection Management defines the Multi-Packet Broadcast and especially the flow-control for destination specific (peer-to-peer) transmission of multi-package messages.

Flow control, in this case, means functions such as broadcast announcement, request to send, clear to send, end of message acknowledgement, connection abort, and more.

The SAE J1939/21 standard fails to mention that multi-packet messages are only supported in PDU1 format (Peer-to-Peer communication). In order to broadcast messages they must be addressed to the global destination address (= 255).

4.5.2.1 Multi-Packet Broadcast

A Broadcast message is per definition not destination specific, meaning the message is broadcasted to all nodes in the network by using the global destination address (= 255).

In order to broadcast a multi-packet message a node must first send a *Broadcast Announce Message* (BAM). A BAM message contains the following components:

- Parameter Group Number of the multi-packet message
- Size of the multi-packet message
- Number of packages

The BAM message allows all receiving nodes (= all nodes interested in the message) to prepare for the reception by allocating the appropriate amount of resources (memory).

The *Broadcast Announce Message* (BAM) is embedded in the Transport Protocol – Connection Management (TP.CM) PGN 60416 and the actual data transfer is handled by using the Data Transfer PGN 60160[48].

Parameter Group Name	**Transport Protocol – Connection Management (TP.CM)**
Parameter Group Number	60416 ($00EC00_{hex}$)
Definition	Used for Communication Management flow-control (e.g. Broadcast Announce Message).
Transmission Rate	According to the Parameter Group Number to be transferred
Data Length	8 bytes
Extended Data Page (R)	0
Data Page	0
PDU Format	236
PDU Specific	Destination Address (= 255 for broadcast)
Default Priority	7
Data Description	(For Broadcast Announce Message only)
Byte	1 - Control Byte = 32
	2,3 – Message Size (Number of bytes)
	4 – Total number of packages
	5 – Reserved (should be filled with FF_{hex})
	6-8 – Parameter Group Number of the multi-packet message (6=LSB, 8=MSB)

[48] The SAE J1939/21 Standard – Chapter 5.10.2.1 Multipacket Broadcast – does not offer any reference regarding the exact structure of the BAM, the BAM PGN, or the Data Transfer PGN 60160.

Parameter Group Name	**Transport Protocol – Data Transfer (TP.DT)**
Parameter Group Number	60160 ($00EB00_{hex}$)
Definition	Data Transfer of Multi-Packet Messages
Transmission Rate	According to the Parameter Group Number to be transferred
Data Length	8 bytes
Extended Data Page (R)	0
Data Page	0
PDU Format	235
PDU Specific	Destination Address
Default Priority	7
Data Description	
Byte	1 – Sequence Number (1 to 255)
	2-8 - Data

The last packet of a multi-packet PGN may require less than eight data bytes. All unused data bytes in the last package are being set to FF_{hex}.

The transport of Multi-Packet Broadcast messages is not regulated by any flow-control functions and thus it is necessary to define timing requirements between the sending of a *Broadcast Announce Message* (BAM) and the Data Transfer PGN. The following picture demonstrates the message sequence and timing requirements for a broadcasted multi-packet message[49].

[49] One would expect that engineers, regardless of their special expertise, are familiar with the unit of time, "ms" or "msec" (milli-seconds). Instead the SAE J1939/21 standard uses mS, which is officially milli-Siemens (electric conductance, equal to inverse Ohm - Ω).

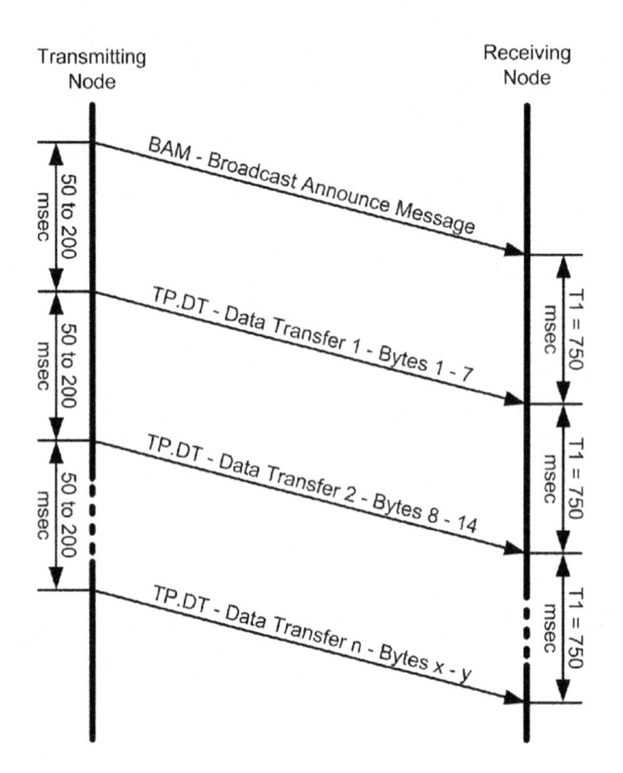

Picture 4.5.2.1.1 Broadcast Data Transfer

As is demonstrated in the picture:

- The message packet frequency will be between 50 to 200 msec.
- A timeout will occur when a time of greater than 750 ms **(T1)** elapsed between two message packets when more packets were expected.

Timeouts will cause a connection closure.

A connection is considered closed when:

- The sender of a data message sends the last Data Transfer package.
- A timeout occurs.

4.5.2.2 Multi-Packet Peer-to-Peer

The communication of destination specific (peer-to-peer) multi-packet message is subject to flow-control. The three basic components of flow control are:

1. **Connection Initialization** – The sender of a message transmits a *Request to Send* message. The receiving node responds with either a *Clear to Send* message or a *Connection Abort* message in case it decides not to establish the connection. A *Connection Abort* as a response to a *Request to Send* message is preferred over a timeout by the connection initiator. The *Clear to Send* message contains the number of packets the receiver is expecting plus the expected sequence number.

2. **Data Transfer** – The sender transmits the Data Transfer PGN after receiving the *Clear to Send* message. Data transfer can be interrupted/stopped by a *Connection Abort* message.

3. **Connection Closure** – The receiver of the message, upon reception of the last message packet, sends an *End of Message ACK* (acknowledgement) message, provided there were no errors during the transmission. Any node, sender or receiver, can send a *Connection Abort* message. The reason of aborting a connection can be a timeout.

A reliable flow-control will must also include timeouts in order to assure proper network function. SAE J1939/21 defines a number of timeouts as listed below and their application is explained in the following.

T_r = 200 ms - Response Time

T_h = 500 ms - Holding Time

T1 = 750 ms

T2 = 1250 ms

T3 = 1250 ms

T4 = 1050 ms

Scenarios for timeout control are:

- A node (regardless whether the node is the receiver or sender of the data message) does not respond within 200 ms **(T_r)** to a data or flow control message[50].
- If a receiving node needs (for any reason) to delay the transmission of data it can send a *Clear to Send* message where the number of packages is set to zero. In cases where the flow must be delayed for a certain time the receiver of a message must repeat the transmission of the *Clear to Send* message every 0.5 seconds **(Th)** to maintain an open connection with the sender of the message. As soon as the receiver is ready to receive the message it must send a regular *Clear to Send* message[51].
- A time of greater than **T1** elapsed between two message packets when more packets were expected.
- A time greater than **T2** elapsed after a *Clear to Send* message without receiving data from the sender of the data.
- A time greater than **T3** elapsed after the last transmitted data packet without receiving a *Clear to Send* or *End of Message Acknowledgment* (ACK) message.
- A time greater than **T4** elapsed after sending a Clear to Send message to delay data transmission without sending another Clear to Send message

Any timeout condition will consequently cause a connection closure.

Other reasons for connection closure are:

- The sender of a data message sends the last Data Transfer package.
- The receiver of a data message receives the last Data Transfer package and a T1 timeout occurs.
- The sender of a data message receives an *End of Message ACK* message.
- Reception of *Connection Abort* message.

The flow control messages, such as *Request to Send*, *Clear to Send*, etc. are embedded in the Transport Protocol – Connection Management (TP.CM) PGN 60416 and the actual data transfer is handled by using the Data Transfer PGN 60160.

[50] The response time is listed, but not explained in SAE J1939/21. It does appear, though, in one of the rare pictures describing data transfer without errors.

[51] This circumstance is not specifically mentioned in SAE J1939/21, but can be derived from other statements made in regards to the Clear to Send message.

Parameter Group Name	**Transport Protocol – Connection Management (TP.CM)**
Parameter Group Number	60416 ($00EC00_{hex}$)
Definition	Used for Communication Management flow-control (e.g. *Request to Send*, *Clear to Send*, etc.).
Transmission Rate	According to the Parameter Group Number to be transferred
Data Length	8 bytes
Extended Data Page (R)	0
Data Page	0
PDU Format	236
PDU Specific	Destination Address (= 255 for broadcast)
Default Priority	7
Data Description	Depending on content of Control Byte – See following description.

TP.CM_RTS	**Connection Mode Request to Send**
	1 - Control Byte = 16
	2,3 – Message Size (Number of bytes)
	4 – Total number of packets
	5 – Max. number of packets in response to CTS. No limit when filled with FF_{hex}.
	6-8 – Parameter Group Number of the multi-packet message (6=LSB, 8=MSB)
TP.CM_CTS	**Connection Mode Clear to Send**
	1 - Control Byte = 17
	2 - Total number of packets (should not exceed byte 5 in RTS)
	3 – Next packet number
	4,5 – Reserved (should be filled with FF_{hex})
	6-8 – Parameter Group Number of the multi-packet message (6=LSB, 8=MSB)
TP.CM_EndOfMsgACK	**End of Message Acknowledgment**
	1 - Control Byte = 19
	2,3 – Message Size (Number of bytes)
	4 – Total number of packages

5 – Reserved (should be filled with FF_{hex})

6-8 – Parameter Group Number of the multi-packet message (6=LSB, 8=MSB)

TP.Conn_Abort

Connection Abort

1 - Control Byte = 255

2 – Connection Abort Reason (See following description)

3-5 – Reserved (should be filled with FFhex)

6-8 – Parameter Group Number of the multi-packet message (6=LSB, 8=MSB)

Control Byte = 32 is reserved for *Broadcast Announce Message*. Control Bytes 0-15, 18, 20-31, 33-254 are reserved by the SAE.

The Connection Abort Reasons can be:
1 – Node is already engaged in another session and cannot maintain another connection.
2 – Node is lacking the necessary resources.
3 – A timeout occurred.
4...250 - Reserved by SAE.
251...255 – Per J1939/71 definitions[52].

The SAE J1939/21 standard is not very specific when it comes to the extended use of the *Clear to Send* message. The following paragraph will explain these features in more detail.

If a receiving node needs (for any reason) to delay the transmission of data it can send a *Clear to Send* message where the number of packages is set to zero. In cases where the flow must be delayed for a certain time the receiver of a message must repeat the transmission of the *Clear to Send* message every 0.5 seconds **(Th)** to maintain an open connection with the

[52] No further references or explanations are offered in the SAE Standards Collection.

sender of the message. As soon as the receiver is ready to receive the message it must send a regular *Clear to Send* message[53].

What SAE J1939/21 fails to mention is that the *Clear to Send* message can be send by the receiver of the data message at any time, either immediately after the reception of a *Request to Send* message or after reception of a data packet, meaning any time during the data transfer.

The data transfer is handled by using the Data Transfer PGN 60160.

Parameter Group Name	**Transport Protocol – Data Transfer (TP.DT)**
Parameter Group Number	60160 ($00EB00_{hex}$)
Definition	Data Transfer of Multi-Packet Messages
Transmission Rate	According to the Parameter Group Number to be transferred
Data Length	8 bytes
Extended Data Page (R)	0
Data Page	0
PDU Format	235
PDU Specific	Destination Address
Default Priority	7
Data Description	
Byte	1 – Sequence Number (1 to 255)
	2-8 – Data

The last packet of a multi-packet PGN may require less than eight data bytes. All unused data bytes in the last package are being set to FF_{hex}.

[53] This circumstance is not specifically mentioned in SAE J1939/21, but can be derived from other statements made in regards to the Clear to Send message.

J1939 Network Management

The SAE J1939 Network Management is defined in SAE J1939/81. This document is in far better condition than, for instance, SAE J1939/21. After all, there is a visible structure. Still, the authors use terms right from the beginning that are then being explained in later chapters. And, yet again, the authors indulge themselves in acronyms, especially CA (Controller Application), which is being excessively used throughout the document.

Network Management under J1939 is primarily represented by the Address Claiming Process. While other higher layer protocols based on Controller Area Network (CAN) do not support dynamic node address assignments per default, the SAE J1939 standard provides this ingeniously designed feature to uniquely identify ECUs and their primary function.

SAE J1939/81 prefers the use of CA (Controller Application) rather than ECU (Electronic Control Unit). In all consequence one ECU can run multiple CAs. Each Controller Application will have one address and associated NAME (See following chapters). The following chapters will continue using the term ECU, which is a synonym for CA[54].

[54] The SAE J1939 document provides an overview of the address claiming procedure and this document, while referring to SAE J1939/81 for more details, uses only the term ECU.

5.1 Address Claiming Procedure Overview

While other higher layer protocols based on CAN do not support dynamic node address assignments per default, the SAE J1939 standard provides yet another ingeniously designed feature to uniquely identify ECUs and their primary function.

The CAN standard in itself does not support node (ECU) addresses, only message IDs, where one node may manage multiple messages. However, the message ID must be hard-coded in the application program. Also, in a standard CANopen network the node address is usually hard-wired or mechanically adjustable (e.g. per dip switch).

Each ECU in a J1939 vehicle network must hold at least one NAME and one address for identification purposes. Single electronic units are allowed, however, to control multiple names and addresses.

The 8-bit ECU address defines the source or destination for messages.

The ECU NAME includes an indication of the ECU's main function performed at the ECU's address. A function instance indicator is added in cases where multiple ECUs with the same main function share the same network.

The J1939 standard allows up to 253 ECUs with the same function to share the same network, where each ECU is identified by their individual address and NAME[55].

[55] This statement was derived from the SAE J1939 standard and is somewhat misleading. J1939 allows a maximum of 30 ECUs, but a maximum of 253 Controller Applications.

SAE J1939 defines a 64 bit NAME, as shown in the picture below, to uniquely identify each ECU in a network.

Arbitrary Address Capable	Industry Group	Vehicle System Instance	Vehical System	Reserved	Function	Function Instance	ECU Instance	Manufacturer Code	Identity Number
1 bit	3 bit	4 bit	7 bit	1 bit	8 bit	5 bit	3 bit	11 bit	21 bit

Picture 5.1.1 J1939 NAME Fields

While the 64 bit NAME is certainly appropriate to uniquely identify nodes (ECUs) and their function in a J1939 network, it will nevertheless necessitate unreasonable resources to maintain standard communications.

In order to provide a more efficient solution, the SAE J1939 Standard defines an address claim procedure[56], where each ECU utilizes an 8 bit address to identify the source of a message or to access (destination address) another ECU in the network. The address claim procedure is designed to assign addresses to ECUs right after the network has been initialized and thus assuring that the assigned address is unique to the ECU. For instance, an engine may be assigned the address 0 while another engine is present, which will be assigned another address (e.g. 1) and instance.

ECUs designed to accept destination specific commands may require multiple addresses, each with their corresponding NAME, in order to distinguish the required action. For instance, the torque from the engine as commanded by the transmission must be separated from the torque commanded by the brake.

[56] The address claim procedure is defined through the SAE J1939/81 Standard and elaborated on in SAE J1939/01.

In addition, the SAE J1939 Standard defines Preferred Addresses[57] to commonly used devices in order to minimize the rate of multiple devices demanding the same address and consequently optimizing the address claim process. ECUs will generally use their assigned Preferred Address immediately after the power up process, but in order to prevent any address claim conflicts, each ECU must first announce which addresses it intends to claim.

The address claim feature considers two possible scenarios:

- **Sending an Address Claimed message**

  This first scenario addresses a standard J1939 network startup. Upon powering up (or when requested), an ECU will send an Address Claimed message into the CAN bus in order to claim an address. All ECUs receiving the address claim will record and verify the newly claimed address with their internal address table[58]. In case of an address conflict, i.e. should two or more ECUs claim the same address, the ECU with the lowest NAME value will succeed and use the address as claimed. The remaining ECUs must claim a different address or stop transmitting to the network.

- **Request for Address Claimed message**

  Requesting an Address Claimed message from all nodes in the network and, as a result, determining addresses claimed by other ECUs, is the necessary procedure for ECUs powering up late for various reasons, but especially transitional ECUs. Such transitional ECUs may be diagnostics tools, service tools, and trailers, etc. This approach allows the ECU to determine and claim an available address or to find out which ECUs are currently on the network.

[57] Preferred Addresses are listed in the SAE J1939 document (Table B2 – B9) and are not part of this book. Please refer to http://www.j1939standardscollection.com for further references.

[58] Keeping an address table is not mandatory to comply with the SAE J1939 Standard, but ECUs must be at least capable to compare newly claimed addresses with their own.

The request for an Address Claimed message supports primarily self-configurable ECUs. The self-configurable addressing feature is not a requirement for all ECUs, i.e. it is optional, but it is highly recommended for those ECUs for which an address claiming conflict is considered to be highly probable.

There are several options to resolve the situation where an address claim conflict is being detected and they depend on the ECU's capabilities. The address claim capability is categorized into two classes:

1. Single Address Capable ECUs
2. Arbitrary Address Capable ECUs

The definition of Address Claiming Capabilities in SAE J1939/81 is somewhat vague, since (according to SAE J1939/81) "the classifications are not necessarily mutually exclusive".

5.1.1 Single Address Capable ECUs

There are several methods by which an ECU can support a Single Address Capability and they are listed in the following. These methods require the intervention through some external process and the range of addresses they claim is limited.

➢ **Self-Configurable ECUs**

A self-configurable ECU provides the ability of dynamically processing and claiming unused addresses, however, from a limited set of ECU addresses. Such ECUs are largely diagnostics tools, service tools, and bridges, etc.

SAE J1939/81 emphasizes that a self-configurable ECU cannot be considered being an Arbitrary Address capable ECU (See following description). The context in which this statement was made suggests that the NAME will have no influence in case a self-configurable ECU encounters an address conflict. This is, however, in contradiction to the definition of the self-configurable ECU.

➢ **Command Configurable ECUs**[59]

Service tools or bridges may request (command) certain ECUs to assume an address that specifically serves the function of the service tool or bridge.

➢ **Service Configurable ECUs**[60]

These are ECUs whose address may be changed manually by service personnel (e.g. DIP switches) or through a service tool. Service tools usually use proprietary techniques and, as a result, the process of commanding a new address may differ from the process as described for Command Configurable ECUs.

➢ **Non-Configurable ECUs**[61]

As the name implies, Non-Configurable ECUs are neither self-configurable nor re-programmable, i.e. their address cannot be changed after they claimed it initially.

5.1.2 Arbitrary Address Capable ECUs

The SAE J1939/81 standard uses a lot of sophisticated, but nevertheless non-descriptive sentences, to explain that an ECU with Arbitrary Address capability is able to re-calculate and re-claim addresses from a broad range in case an address claim conflict occurred.

---

[59] The description in SAE J1939 about the specifics of the address commanding process is ambiguous. In all consequence there could be two ECUs that need to change their address: 1. the ECU that receives the command and 2. the ECU that already owns the commanded address. This scenario has not been addressed in this document. The author has used an abbreviated description of Command Configurable ECUs to eliminate the potential of presenting incorrect information.

[60] Yet again, this section in the SAE J1939 document is ambiguous in regards to the service tools. The author has used his best judgment to describe this type of ECU. The section also contains an apparent typo that turns the interpretation of the topic into a guessing game.

[61] And yet again, this section in the SAE J1939 document is ambiguous and does not provide any supporting details.

5.2 ECU NAME and Addresses

Each ECU in a J1939 vehicle network must hold at least one NAME and one address for identification purposes. Single electronic units are allowed, however, to control multiple names and addresses.

The 8-bit ECU address defines the source or destination for messages. The ECU NAME includes an indication of the ECU's main function performed at the ECU's address, but can also influence the address claiming process in case of an address claiming conflict.

5.2.1 NAME

SAE J1939/81 defines a 64 bit NAME, as shown in the picture below, to uniquely identify each ECU in a network.

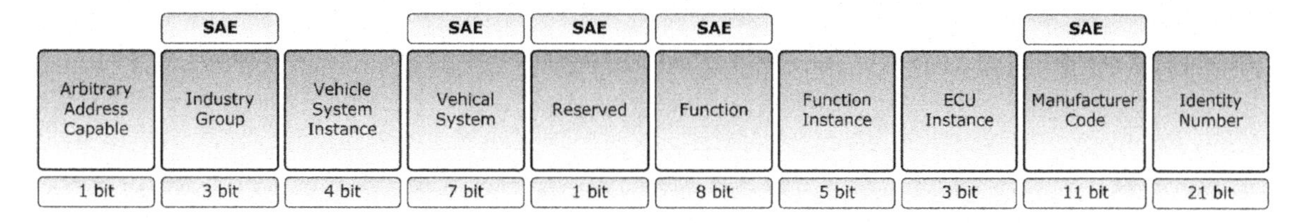

Picture 5.2.1.1 J1939 NAME Fields

The NAME field contains 10 entries of which five are assigned by the SAE[62]. The remaining five fields are either derived from the network characteristics or are manufacturer specific.

The NAME field is not only important to uniquely identify the Controller Application; it is also being used during the address claim process when two or more ECUS are attempting to claim the same address (See also chapter *Network Management Procedure*). In such a case the ECU with a NAME of higher priority (= lower numeric value) will claim the address.

[62] SAE J1939/81 claims that six fields are assigned by the SAE committee, but fails to provide evidence.

The NAME, which is always associated with an address, is usually configured during the initial vehicle/machine configuration (i.e. the final vehicle assembly) or when the ECU is installed.

5.2.1.1 NAME Field: Arbitrary Address Capable

Indicates whether or not the ECU is Arbitrary Address capable (1=yes, 0=no). If the ECU is Arbitrary Address capable and encounters an address conflict the ECU with a NAME of higher priority (= lower numeric value) will claim the address; the other one will claim another free address.

SAE J1939/81 emphasizes that a self-configurable ECU cannot be considered being an Arbitrary Address capable ECU. The context in which this statement was made suggests that the NAME will have no influence in case a self-configurable ECU encounters an address conflict. This is, however, in contradiction to the definition of the self-configurable ECU.

5.2.1.2 NAME Field: Industry Group

Industry Group codes are associated with particular industries, which can be, for instance, on-highway equipment, agricultural equipment etc. The 3 bit code is assigned by the SAE and definition can be found in the SAE J1939 standard[63].

[63] SAE J1939/81 refers to a reference in the SAE J1939 document that does not exist.

The following table shows the currently assigned Industry Group codes.

Industry Group #	Industry
0	Global, applies to all industries
1	On-Highway Equipment
2	Agricultural and Forestry Equipment
3	Construction Equipment
4	Marine
5	Industrial-Process Control-Stationary (Generator Sets)
6, 7	Reserved

Table 5.2.1.2.1: Industry Group Codes

5.2.1.3 NAME Field: Vehicle System Instance

The Vehicle System Instance works in combination with the next field, Vehicle System. A J1939 network may accommodate several ECUs of the same Vehicle System. The 4 bit long Vehicle System Instance assigns a number to each instance of the Vehicle System (0 to 15).

5.2.1.4 NAME Field: Vehicle System

This 7 bit field is defined and assigned by the SAE. Definitions of the Vehicle System are found in the SAE J1939 standard. Vehicle Systems are closely associated with Industry Groups and they can be, for instance, "tractor" in the "Common" industry, "trailer" in the "On-Highway" industry group or "planter" in "Agricultural Equipment".

5.2.1.5 NAME Field: Reserved

The Reserved filed is always set to zero. It is reserved for future definition by the SAE.

5.2.1.6 NAME Field: Function

The Function field is defined and assigned by the SAE. The range of the field is from 0 to 255, but not all values have been assigned.

The interpretation of codes equal to or greater than 128 depends on the Industry Group entry. For instance, a Function code of 133 means "Product Flow" in the "Agricultural and Forestry Equipment" Industry Group. If the Industry Group is "Construction Equipment" the Function is assigned to be "Land Leveling System Display".

The Function code does not depend on any other field in cases where it is less than 128 (0 to 127).

5.2.1.7 NAME Field: Function Instance

The Function Instance works in combination with the Function field. A J1939 network may accommodate several ECUs (Controller Applications) with the same Function. The 5 bit long Function Instance assigns a number to each instance of the Function, where 0 is assigned to first instance.

5.2.1.8 NAME Field: ECU Instance

A J1939 network may accommodate several ECUs (Controller Applications) with the same function. For instance, a single vehicle may accommodate two identical ECUs where the first measures the road speed of the vehicle, while the second one measures the speed of the attached trailer.

The 3 bit long ECU Instance assigns a number to each instance of the ECU, where 0 is assigned to first instance.

5.2.1.9 NAME Field: Manufacturer Code

The 11 bit Manufacturer code is assigned by the SAE and it indicates which manufacturer produced this particular ECU. The Manufacturer Code is defined in the SAE J1939 document.

5.2.1.10 NAME Field: Identity Number

The 21 bit Identity Number is assigned by the manufacturer of the ECU and should be used to guarantee unique NAMEs within a product line. The manufacturer is also allowed to add further information to the Identity Number such as, for instance, serial number, date of manufacture, etc.

5.2.1.11 NAME Field Dependencies

The following picture demonstrates the dependencies inside the NAME field according to the definition of all NAME fields.

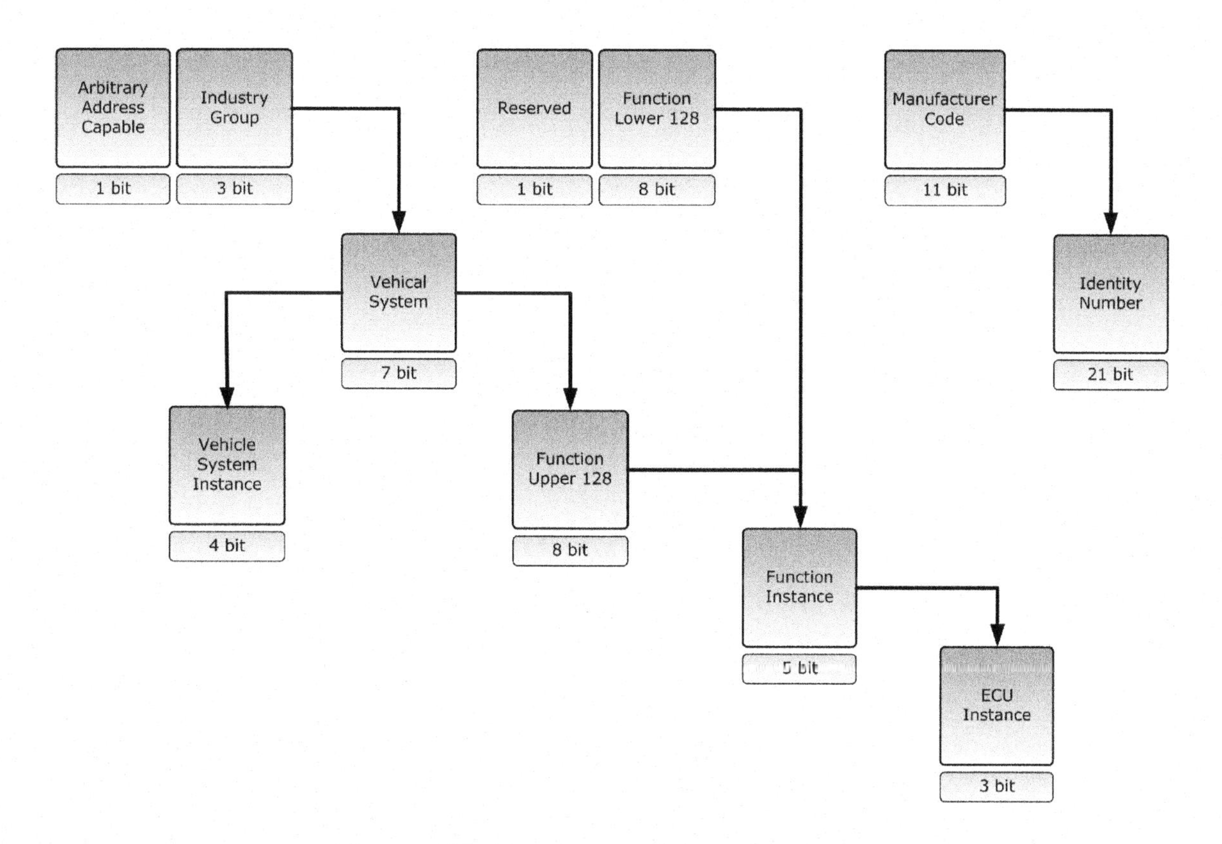

Picture 5.2.11.1 Name Field Dependencies

5.2.2 Addresses

The J1939 protocol utilizes an 8 bit device (ECU) address, which, theoretically, would allow the operation of 256 nodes in the same network. It can only be assumed that the SAE was trying to keep the bus traffic on a low level by restricting the maximum number of nodes to 30. Elaborating comments on this restrictions may be embedded somewhere in the standard.

In all consequence the ECU address is really a Controller Application address in a situation where each ECU may accommodate several Controller Applications. The 253 addresses (Address 254 is reserved for Network Management, Address 255 is used for global addressing – Refer to following chapters) are assigned (claimed) for the Controller Applications, not the actual ECU.

The following picture shows an example, where, for instance, ECU A accommodates three controller applications.

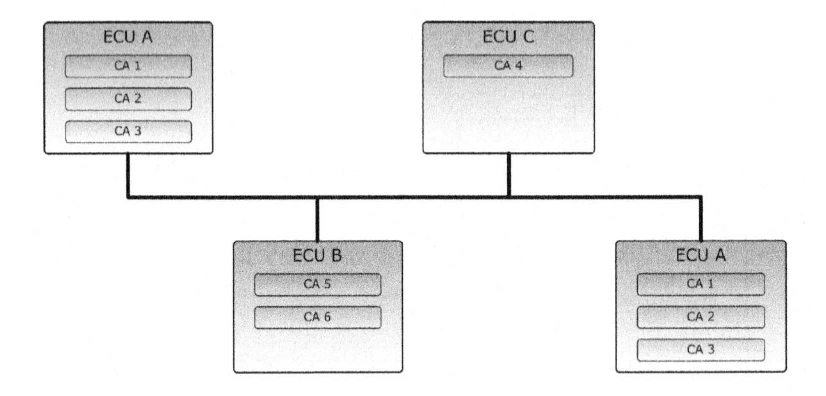

Picture 5.2.2.1 Sample J1939 Network

The picture also demonstrates that ECUs of the same function (ECU A) can co-exist in a J1939 network without address collision. J1939 features a very ingenious feature, the Address Claim procedure which automatically assigns addresses to each Controller Application. In case of an Address Claim conflict, the Controller Applications are able to claim another free address.

5.2.2.1 Preferred Addresses

In order to optimize the address claim procedure (Refer to chapter *Address Claim Procedure*) the SAE J1939 standard lists a number of predefined (preferred) addresses. A preferred address is closely associated with the function of the ECU and it must be reflected in the NAME field (Function). An ECU (or Controller Application) will attempt to claim this assigned address first. Depending on the ECU's addressing capability it must attempt to claim another address in case the original preferred address is already taken.

The range of preferred addresses and their interpretation depends on the Industry Group as demonstrated in the following[64].

Industry Group	**Preferred Address Range**
Global (Applies to all industry groups)	0 – 84 Assigned 85 – 127 Reserved 248, 252 - 255 Reserved
Industry Group #1 – On-Highway Equipment	128 – 160 Dynamic 161 – 247 Assigned
Industry Group #2 – Agricultural and Forestry Equipment	128 – 207 Dynamic 208 – 247 Reserved
Industry Group #3 – Construction Equipment	128 - 207 Dynamic 208 – 247 Reserved
Industry Group #4 – Marine Equipment	128 – 207 Dynamic 208 – 247 Reserved
Industry Group #5 – Industrial, Process Control, Stationary Equipment	128 – 207 Dynamic 208 – 247 Reserved

Table 5.2.2.1.1 Preferred Address Range

[64] Based on SAE J1939 – REV. FEB2007.

Note: "Dynamic" indicates that these addresses are available for self-configurable ECUs in case of address claim conflicts.

5.2.2.2 NULL Address

The address 254, the so-called NULL Address, is reserved for network management (See chapter *Network Management Procedure*) and it is used for the *Cannot Claim Source Address* message.

5.2.2.3 Global Address

The Global Address (255) is exclusively used as a destination address in order to support message broadcasting (sending a message to all network nodes).

5.3 Network Management Procedure

According to SAE J1939/81 network management procedures are used to "collectively manage the network". The chapters on network management have no logical structure (Again, explaining the function of an automobile, starting with the details of the fuel injection system); they explain the address claim messages first in detail and then follow up with the actual procedure.

In all consequence the network management is all about the Address Claim procedure and this procedure utilizes three messages and their PGNs:

- Request Message (PGN 59904)
- Address Claimed / Cannot Claim (PGN 60928)
- Commanded Address (PGN 65240)

5.3.1 Address Claim Procedure

The address claim feature considers two possible scenarios:

➢ **Sending an Address Claimed message**

This first scenario addresses a standard J1939 network startup. Upon powering up (or when requested), an ECU will send an Address Claimed message into the CAN bus in order to claim an address. All ECUs receiving the address claim will record and verify the newly claimed address with their internal address table[65]. In case of an address conflict, i.e. should two or more ECUs claim the same address, the ECU with the lowest NAME value will succeed and use the address as claimed. The remaining ECUs must claim a different address or stop transmitting to the network.

➢ **Request for Address Claimed message**

Requesting an Address Claimed message from all nodes in the network and, as a result, determining addresses claimed by other ECUs, is the necessary procedure for ECUs powering up late for various reasons, but especially transitional ECUs. Such transitional ECUs may be diagnostics tools, service tools, and trailers, etc. This approach allows the ECU to determine and claim an available address or to find out which ECUs are currently on the network.

The request for an Address Claimed message supports primarily self-configurable ECUs. The self-configurable addressing feature is not a requirement for all ECUs, i.e. it is optional, but it is highly recommended for those ECUs for which an address claiming conflict is considered to be highly probable.

After completing their Power On Self Test (POST) ECUs (Controller Applications) claiming addresses in the 0 to 127 or 248 - 253 range may initiate their regular network activities

[65] Keeping an address table is not mandatory to comply with the SAE J1939 Standard, but ECUs must be at least capable to compare newly claimed addresses with their own.

immediately, while other ECUs should not begin until a time of 250 ms after claiming an address. This allows competing claims to be resolved before the address is being used.

In the event that two ECUs attempt to claim the same address, the ECU with the lowest NAME value will succeed and use the address as claimed. The remaining ECUs must claim a different address by sending another *Address Claimed* message containing a different address or send a *Cannot Claim Address* message.

The destination address for an address claim is always the global address (255) in order to address all nodes in the network.

A node, that has not yet claimed an address, must use the NULL address (254) as the source address when sending a *Request for Address Claimed* message.

The following picture demonstrates two possible address claim scenarios.

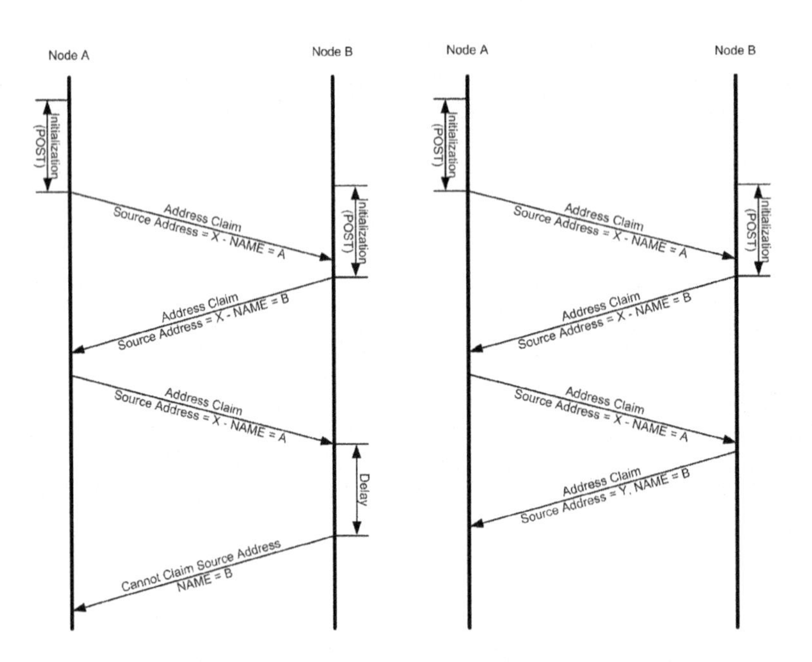

Picture 5.3.1.1 Address Claim Procedure

Both scenarios basically show the same: Two nodes, A and B, are claiming the same address; node A has a NAME of higher priority. Node B, in the left scenario, is not a self-configurable ECU, meaning it requires external actions to claim another address.

In the right scenario, node B is a self-configurable ECU and eventually will claim another source address.

The following describes the steps in detail:

- Node A starts initialization and Power-On Self Test (POST) some time ahead of node B.
- While node B is going through initialization and POST, node A sends out it address claim message.
- Node B, after having finished initialization and POST, attempts to claim the same source address as node A.
- In response node A, having determined that its NAME has higher priority, resends the address claim message.
- Node B receives the address claim message, determines that node A's name has higher priority.
- In the left scenario, node B sends a *Cannot Claim* message. In the right scenario it claims another address by sending another *Address Claim* message.

It is somewhat troubling that SAE J1939/81 allows (and describes) a situation during the address claim process where two CAN nodes with identical message IDs can access the bus at the same time. This scenario is, if the CAN standard has any significance at all, not legitimate. SAE J1939/81 recommends a procedure to solve this situation that can only be considered peculiar.

The scenario as mentioned may happen when two ECUs try to claim the same source address at exactly the same time. The following picture demonstrates this possibility.

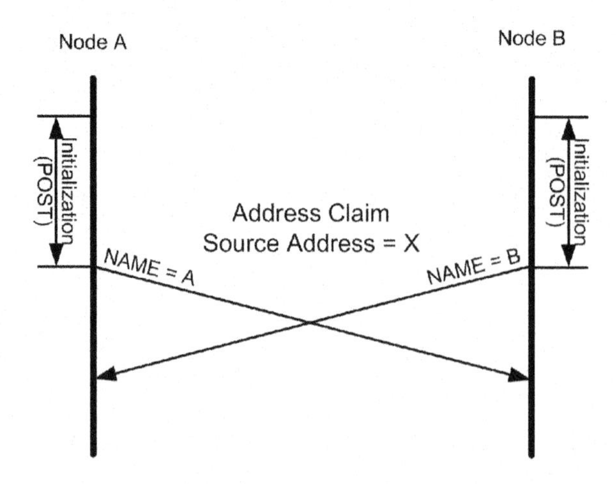

Picture 5.3.1.2 J1939 Message Collision

In this example, both nodes, A and B, are powered up at the same time, their Power-On Self Test (POST) takes the same time and both nodes send out their Address Claim message at the same time. The probability for such a situation taking place may be extremely low, but the consequences can be serious.

The CAN standard does not allow a situation where two nodes use the same message ID. This scenario will create a CAN error frame and both nodes will attempt to claim the same source address repeatedly, until both nodes switch into the Error Passive mode which in turn will limit their network communication capabilities. Experience has shown that both nodes will eventually return to normal operation, but the time it takes is first of all unpredictable and proper network function cannot be guaranteed.

SAE J1939/81 recommends an *Address Claim Bus Collision Management*. The procedure is to check the occurrence of an error frame and to use different delays in both ECUs before they claim their address again. The transmit delay should be calculated by producing a "pseudo-random" delay value between 0 and 255.

This method can only be considered an unprofessional approach, since it provides a solution to a violation of the CAN standard.

SAE J1939/81 explains that, during this process, both nodes would go eventually into BUS OFF mode. However, this statement is not backed up by either the CAN standard (which categorically does not allow two messages with the same ID) or empirical tests.

Tests (not accomplished by the SAE) have shown that the two competing nodes will go into Error Passive mode and both nodes will eventually return to the regular Error Passive mode. However, the time until both nodes return to regular activities is unpredictable and so are the consequences for the application.

5.3.2 Address Management Messages

The network management messages have the same characteristics as all other J1939 messages. The messages are:

Message	**PGN**	**PF**	**PS**	**SA**	**Data Length**	**Data**
Request for Address Claimed	59904	234	DA	SA[1)]	3 bytes	PGN 60928
Address Claimed	60928	238	255	SA	8 bytes	NAME
Cannot Claim Source Address	60928	238	255	254	8 bytes	NAME
Commanded Address	65240	254	216	SA	9[2)]	NAME, new SA

Table 5.3.2.1 Address Management Messages

1) In case no address has been claimed as of yet the source address could be set to 254.
2) The commanded address, since it is longer than 8 bytes, is sent using the Transport Protocol as described in chapter *Transport Protocol*.

5.3.2.1 Request for Address Claimed

The Request for Address Claimed message (PGN 59904) is identical to the Request message type as described in SAE J1939/21 and chapter *Parameter Group Numbers* in this book.

Parameter Group Name	**Request**
Parameter Group Number	59904 (00EA00$_{hex}$)
Definition	Requests a Parameter Group from a single device or all devices in the network.
Transmission Rate	User defined (no more 2 to 3 times a second is recommended)
Data Length	3 bytes (CAN DLC = 3)
Extended Data Page (R)	0
Data Page	0
PDU Format	234
PDU Specific	Destination Address (Global or Peer-to-Peer)
Default Priority	6
Data Description	Requested Parameter Group Number = PGN 60928

The *Request for Address Claimed* message is used to request the sending of an *Address Claimed* message from either a particular node in the network or from all nodes (use of global destination address = 255). The Address Claimed message (as described in the following chapter) will provide the requested information, i.e. address and NAME of the responding node(s).

The purpose of sending such a request may be for several reasons, for instance:

- A node is checking whether or not it can claim a certain address.
- A node is checking for the existence of another node (Controller Application) with a certain function.

The response to a Request for Address Claimed message can be multiple:

- Any addressed node that has already claimed an address will respond with an *Address Claimed* message.
- Any addressed node that was unable to claim an address will respond with a *Cannot Claim Address* message.
- Any addressed node that has not yet claimed an address should do so by responding with their own *Address Claimed* message where the source address is set to NULL (254).
- A node sending the Request for Address Claimed message should respond to its own request in case the global destination address (255) was used.

5.3.2.2 Address Claimed / Cannot Claim

The *Address Claimed* message is used either, as the name indicates, to claim a message or to respond to a *Request for Address Claimed* message.

The following rules apply:

- The *Address Claimed* message, for the purpose of claiming an address, should always be addressed to the global address (255).
- The *Address Claimed* message, for the purpose of claiming an address, should be sent during the initialization of the network or as soon as the node is connecting to a running network.
- As soon as a node has successfully claimed an address, it may begin with regular network activities, i.e. sending messages or responding to messages.
- If a node (Controller Application) receives an *Address Claimed* message it should first compare the claimed address with its own. If the addresses are identical, the node should compare its own NAME to the NAME of the claiming node. In case its own NAME has a higher priority (lower numeric value) it will then transmit an *Address Claimed* message containing its NAME and address. If its own NAME is of a lower priority the node, depending on its capabilities, should either send a *Cannot Claim Address* message or claim another address.
- In case a node loses its address through the previously described procedure and was also in the process of sending a Transport Protocol message (see chapter *Transport Protocol Functions*) it should cease the transmission immediately, however, without sending a Transport Protocol Abort message. The receiver of the Transport Protocol message will detect the interruption through the corresponding timeout process.

Parameter Group Name	**Request**
Parameter Group Number	60928 ($00EE00_{hex}$)
Definition	Address Claimed Message
Transmission Rate	As required.
Data Length	8 bytes (CAN DLC = 8)
Extended Data Page (R)	0
Data Page	0
PDU Format	238
PDU Specific	255 = Global Destination Address
Default Priority	6
Data Description	**NAME of Controller Application**
Byte 1	Bits 8-1: LSB of Identity Field
Byte 2	Bits 8-1: 2nd byte of Identity Field
Byte 3	Bits 8-6: LSB of Manufacturer Code
	Bits 5-1: MSB of Identity Field
Byte 4	Bits 8-1: MSB of Manufacturer Code
Byte 5	Bits 8-4: Function Instance
	Bits 3-1: ECU Instance
Byte 6	Bits 8-1: Function
Byte 7	Bits 8-2: Vehicle System
	Bit 1: Reserved
Byte 8	Bit 8: Arbitrary Address Capable
	Bits 7-5: Industry Group
	Bits 4-1: Vehicle System Instance

The *Cannot Claim Address* message has the same format as the *Address Claimed* message, but it uses the NULL address (254) as the source address.

The following rules apply for the *Cannot Claim Address* message:

- As the name implies, a node without arbitrary addressing capabilities will send a *Cannot Claim Address* message when it is unable to claim the preferred address.
- A node with arbitrary addressing capabilities will send a *Cannot Claim Address* message when no addresses are available in the network.

- If the *Cannot Claim Address* message is a response to a *Request for Address Claimed* message, the node should apply a "pseudo-random" delay[66] of 0 to 153 ms should be applied before sending the response. This will help prevent the possibility of a bus error, which will occur when two nodes send the same message with identical message ID.

SAE J1939/81 is concerned that the occurrence of such an error condition will consume "a large number of bit times on the bus". The time for transmitting a CAN message with a 29-Bit ID followed by an error frame will be well under 1 ms, which pales in comparison to a possible 153 ms delay.

5.3.2.3 Commanded Address

The *Commanded Address* message is used to instruct an ECU (Controller Application) with a certain NAME to assume the address as commanded. This procedure is useful, for instance, for bridges, diagnostics and scan tools.

A node (ECU receiving the *Commanded Address* message can respond in two ways:

- The node accepts the new address and starts the address claim procedure[67].
- The node ignores the message and sends no response.

The *Commanded Address* message requires 9 bytes of data which necessitates using the BAM (Broadcast Announce Mode) of the Transport Protocol (as described in SAE J1939/21 or chapter *Transport Protocol Functions* in this book).

[66] The SAE J1939/81 uses a wrong reference that was supposed to explain the method of calculating the delay.

[67] SAE J1939/81 does not provide any insights on a situation where the new address is already claimed by another Controller Application with a NAME of higher priority than the commanded node.

Parameter Group Name	**Request**
Parameter Group Number	60928 (00EE00$_{hex}$)
Definition	Address Claimed Message
Transmission Rate	As required.
Data Length	8 bytes (CAN DLC = 8)
Extended Data Page (R)	0
Data Page	0
PDU Format	238
PDU Specific	255 = Global Destination Address
Default Priority	6
Data Description	**NAME of Commanded Address Target**
Byte 1	Bits 8-1: LSB of Identity Field
Byte 2	Bits 8-1: 2nd byte of Identity Field
Byte 3	Bits 8-6: LSB of Manufacturer Code
	Bits 5-1: MSB of Identity Field
Byte 4	Bits 8-1: MSB of Manufacturer Code
Byte 5	Bits 8-4: Function Instance
	Bits 3-1: ECU Instance
Byte 6	Bits 8-1: Function
Byte 7	Bits 8-2: Vehicle System
	Bit 1: Reserved
Byte 8	Bit 8: Arbitrary Address Capable
	Bits 7-5: Industry Group
	Bits 4-1: Vehicle System Instance
Data Description	**Address Assignment**
Byte 9	Bits 8-1: New Source Address

Appendix A – Web Site References

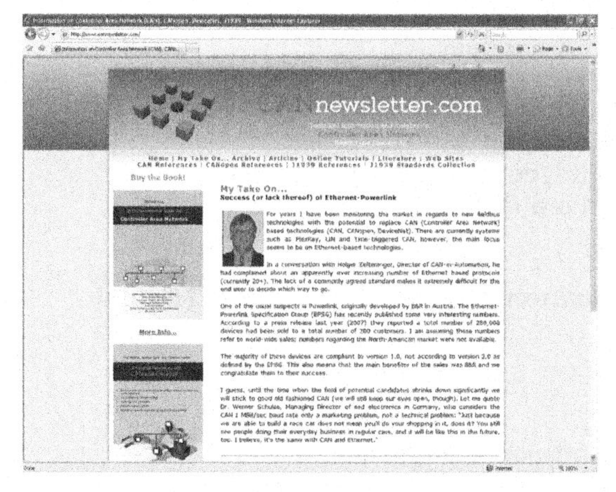

http://www.cannewsletter.com

CANNewsletter.com provides vast information on all aspects of Controller Area Network including CANopen and J1939. The web site contains all kinds of information beyond the standards including many articles and links to CAN, CANopen and J1939 seminars and literature.

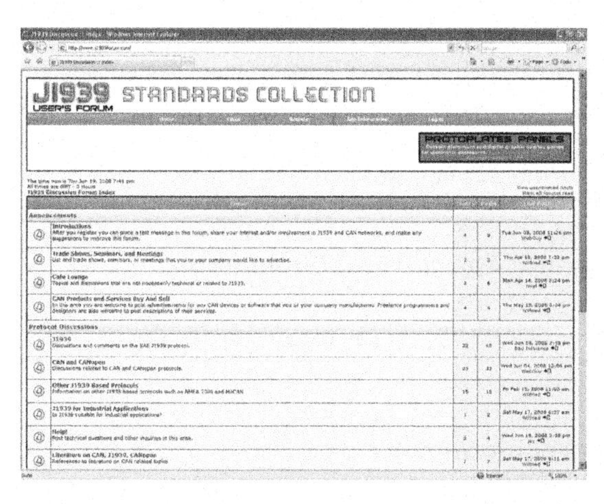

http://www.j1939forum.com

J1939Forum.com is the Online meeting place where to find additional information on SAE J1939 and get help with issues related to SAE J1939.

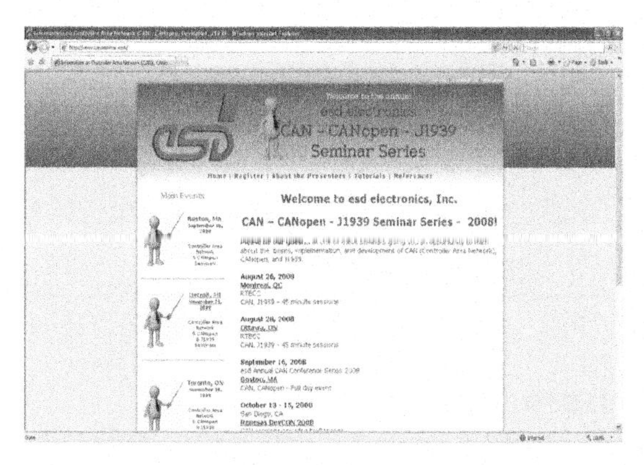

http://www.canseminar.com

Check out this web site, CANSeminar.com, for seminars on CAN, CANopen, or J1939 in your area.

Appendix B – Literature References

1. **A Comprehensible Guide to Controller Area Network**
   by Wilfried Voss
   Published 2005 by Copperhill Technologies Corporation
   http://www.copperhillmedia.com

2. **Embedded Networking with CAN and CANopen**
   by Olaf Pfeiffer, Andrew Ayre and Christian Keydel
   Republished by Copperhill Technologies Corporation
   http://www.copperhillmedia.com

3. **SAE Truck and Bus Control & Communications Network Standards Manual**
   2007 Edition
   HS-1939
   SAE International

4. **Application of J1939 Networks in Agricultural Equipment**
   by Marvin L. Stone and Mark Zachos
   Oklahoma State University - Dearborn Group
   Stillwater, Oklahoma - Farmington Hills, Michigan

5. **J1939-based application profiles**
   by Holger Zeltwanger - CiA
   Reference found at: http://www.can-expo.ru/files/J1939.pdf

6. **High Speed Networking in Construction and Agricultural Equipment**
   by Marvin L. Stone
   Department of Biosystems and Agricultural Engineering
   Oklahoma State University

7. **CAN und J1939 unter extremen Einsatzbedingungen**
   Elektronik automotive 5.2006

8. **A Framework for Developing SAE J1939 Devices**
   by Joachim Stolberg, IXXAT Automation
   March 21, 2007

9. **Introduction to J1939**
   Application Note AN-ION-1-3100
   by Markus Junger
   2004 - Vector Informatik GmbH

10. **Dynamic Address Configuration in SAE J1939**
    by Marvin L. Stone
    Department of Biosystems and Agricultural Engineering

Oklahoma State University

11. **On calculating guaranteed message response times on the SAE J1939 bus**
    by Roger Johansson and Jan Torin
    Chalmers Lindholmen University College
    Goeteborg, Sweden 2002

12. **J1939 C Library for CAN-Enabled PICmicro® Microcontrollers**
    by Kim Otten, Kim Otten Software Consulting
    and Caio Gübel, Microchip Technology Inc.
    Source: http://ww1.microchip.com/downloads/en/AppNotes/00930a.pdf

13. **NMEA 2000 Network Design**
    Jack Rabbit Marine
    Source: http://www.jackrabbitmarine.com/files/NMEA2K_Network_Design_v2.pdf

14. **Q&A - What is SAE J1939?**
    Axiomatic Global Electronic Solutions
    July 6, 2006 – Application Note

Appendix C – Picture Index

Appendix D - Abbreviations

ABS	Antilock Braking System
ACK	Acknowledgement
BAM	Broadcast Announce Message
CA	Controller Application
CAN	Controller Area Network
CM	Connection Management
CRC	Cyclic Redundancy Check
CTS	Clear to Send
DA	Destination Address
DLC	Data Length Code
DP	Data Page
DT	Data Transfer
ECU	Electronic Control Unit
EDP	Extended Data Page
EOF	End of Frame
GE	Group Extension
ID	Identifier
IDE	Identifier Extension Bit
LLC	Logical Link Control
LSB	Least Significant Bit or Byte
MSB	Most Significant Bit or Byte
NA	Not Allowed
NACK	Negative Acknowledgement
P	Priority
PDU	Protocol Data Unit
PF	PDU Format
PG	Parameter Group
PGN	Parameter Group Number
PS	PDU Specific
RTR	Remote Transmission Request
RTS	Request to Send
SA	Source Address
SAE	Society of Automotive Engineers
SOF	Start of Frame
SPN	Suspect Parameter Number
SRR	Substitute Remote Request
TP	Transport Protocol

Index

F

G

H

I

J

M

N

O

CPSIA information can be obtained at www.ICGtesting.com
Printed in the USA
BVOW09s1406200715

409533BV00015B/120/P